FLORIDA STATE
UNIVERSITY LIBRARIES

DEC 2 1996

TALLAHASSEE, FLORIDA

Darwinian Archaeologies

INTERDISCIPLINARY CONTRIBUTIONS TO ARCHAEOLOGY

Series Editor: Michael Jochim, *University of California, Santa Barbara*
Founding Editor: Roy S. Dickens, Jr., *Late of University of North Carolina, Chapel Hill*

Current Volumes in This Series:

THE ARCHAEOLOGY OF WEALTH
Consumer Behavior in English America
James G. Gibb

CASE STUDIES IN ENVIRONMENTAL ARCHAEOLOGY
Edited by Elizabeth J. Reitz, Lee A. Newsom, and Sylvia J. Scudder

CHESAPEAKE PREHISTORY
Old Traditions, New Directions
Richard J. Dent, Jr.

DARWINIAN ARCHAEOLOGIES
Edited by Herbert Donald Graham Maschner

DIVERSITY AND COMPLEXITY IN PREHISTORIC MARITIME SOCIETIES
A Gulf of Maine Perspective
Bruce J. Bourque

HUMANS AT THE END OF THE ICE AGE
The Archaeology of the Pleistocene–Holocene Transition
Edited by Lawrence Guy Straus, Berit Valentin Eriksen, Jon M. Erlandson, and David R. Yesner

PREHISTORIC CULTURAL ECOLOGY AND EVOLUTION
Insights from Southern Jordan
Donald O. Henry

REGIONAL APPROACHES TO MORTUARY ANALYSIS
Edited by Lane Anderson Beck

STATISTICS FOR ARCHAEOLOGISTS
A Commonsense Approach
Robert D. Drennan

STONE TOOLS
Theoretical Insights into Human Prehistory
Edited by George H. Odell

STYLE, SOCIETY, AND PERSON
Archaeological and Ethnological Perspectives
Edited by Christopher Carr and Jill E. Neitzel

A Chronological Listing of Volumes in this series appears at the back of this volume.

A Continuation Order Plan is available for this series. A continuation order will bring delivery of each new volume immediately upon publication. Volumes are billed only upon actual shipment. For further information please contact the publisher.

Darwinian Archaeologies

Edited by

HERBERT DONALD GRAHAM MASCHNER

University of Wisconsin–Madison
Madison, Wisconsin

Foreword by

STEPHEN SHENNAN

Institute of Archaeology
University College London
London, England

PLENUM PRESS • NEW YORK AND LONDON

Library of Congress Cataloging-in-Publication Data

Darwinian archaeologies / edited by Herbert Donald Graham Maschner.
    p.   cm. -- (Interdisciplinary contributions to archaeology)
  Includes bibliographical references and index.
  ISBN 0-306-45328-2
  1. Social archaeology. 2. Social Darwinism. 3. Social evolution.
I. Maschner, Herbert D. G.   II. Series.
CC72.4.D37 1996
930.1--dc20   96-17069
                                                               CIP

ISBN 0-306-45328-2

© 1996 Plenum Press, New York
A Division of Plenum Publishing Corporation
233 Spring Street, New York, N. Y. 10013

10 9 8 7 6 5 4 3 2 1

All rights reserved

No part of this book may be reproduced, stored in a retrieval system, or transmitted in any form or by any means, electronic, mechanical, photocopying, microfilming, recording, or otherwise, without written permission from the Publisher

Printed in the United States of America

In memory of Ben Cullen
February 10, 1964 to December 29, 1995

Contributors

Alysia L. Abbott • Department of Anthropology, University of New Mexico, Albuquerque, New Mexico 87131
Kenneth M. Ames • Department of Anthropology, Portland State University, Portland, Oregon 97207
Robert L. Bettinger • Department of Anthropology, University of California, Davis, California 95616
Robert Boyd • Department of Anthropology, University of California, Los Angeles, California 90024
Ben R. S. Cullen[†] • Sherwood House, St Dogmael's, Cardigan, West Wales SA43 3LF, United Kingdom
Roland Fletcher • Department of Archaeology, School of Archaeology, Classics, and Ancient History, The University of Sydney, New South Wales 2006, Australia
Paul Graves-Brown • Department of Psychology, University of Southampton, Highfield, Southampton SO17 IBJ, United Kingdom
George T. Jones • Department of Anthropology, Hamilton College, Clinton, New York 13323
Robert D. Leonard • Department of Anthropology, University of New Mexico, Albuquerque, New Mexico 87131
Herbert D. G. Maschner • Department of Anthropology, University of Wisconsin, 1180 Observatory Drive, 5240 Social Science, Madison, Wisconsin 53706
Steven Mithen • Department of Archaeology, University of Reading, Whiteknights, P.O. Box 218, Reading RG6 2AA, United Kingdom
Michael J. O'Brien • Department of Anthropology, University of Missouri, 200 Swallow Hall, Columbia, Missouri 65211

[†] Deceased.

John Q. Patton • Department of Anthropology, University of California, Santa Barbara, California 93106

Peter J. Richerson • Division of Environmental Studies, University of California, Davis, California 95616

James Steele • Department of Archaeology, University of Southampton, Highfield, Southampton SO17 IBJ, United Kingdom

Foreword

Just over 20 years ago the publication of two books indicated the reemergence of Darwinian ideas on the public stage. E.O. Wilson's *Sociobiology: The New Synthesis* and Richard Dawkins' *The Selfish Gene*, spelt out and developed the implications of ideas that had been quietly revolutionizing biology for some time. Most controversial of all, needless to say, was the suggestion that such ideas had implications for human behavior in general and social behavior in particular. Nowhere was the outcry greater than in the field of anthropology, for anthropologists saw themselves as the witnesses and defenders of human diversity and plasticity in the face of what they regarded as a biological determinism supporting a right-wing racist and sexist political agenda. Indeed, how could a discipline inheriting the social and cultural determinisms of Boas, Whorf, and Durkheim do anything else? Life for those who ventured to challenge this orthodoxy was not always easy.

In the mid-1990s such views are still widely held and these two strands of anthropology have tended to go their own way, happily not talking to one another. Nevertheless, in the intervening years Darwinian ideas have gradually begun to encroach on the cultural landscape in variety of ways, and topics that had not been linked together since the mid-19th century have once again come to be seen as connected. Modern genetics turns out to be of great significance in understanding the history of humanity. Historical linguistics— one of the disciplines that influenced Darwin's own ideas—is itself increasingly looking toward Darwinian models, while the links between the genetic and the linguistic history of populations are also being explored. There are Darwinian perspectives on how the mind/brain develops, while in the fields of philosophy and psychology it has been postulated that "universal Darwinism" is the key to understanding the nature of human knowledge. Darwinian concepts of variation and selection have even been applied in provocative and stimulating ways to literary history and criticism. Darwinism, it appears, is increasingly in tune with the Zeitgeist.

How does archaeology fit in with all this? The answer is a complex one. The cultural-historical archaeology of the first half of this century can certainly be seen in retrospect as fulfilling aspects of a Darwinian program even though this is not how it was perceived at the time. The process of documenting the history of archaeological cultures and their origins represented the demonstration of descent with modification in the field of human culture, even if the mechanisms responsible for this were inadequately and inaccurately characterized. The processual archaeology that succeeded it in the Anglo-American sphere of influence certainly thought of itself as evolutionary because of its concern with adaptation. However, it was shot through with the fallacies of group functionalism, as Maschner and Mithen point out, and it was Lamarckian rather than Darwinian in character because change was conceived in terms of the environment impressing itself on human populations, as opposed to internally and historically generated variation within human populations interacting with, and being selected by, aspects of the environment.

This functionalist approach probably remains the disciplinary orthodoxy in many parts of the world, but in Britain and elsewhere it has been superseded in this role by so-called "post-processual" archaeology whose hallmarks include a concern with humanistic interpretation rather than scientific explanation, with meaning rather than function, with individuals and the contingencies of history rather than the formulation of generalizations, and with the social context of archaeology and contextual influences on its ideas as opposed to their internal coherence and their relationship to evidence.

Nevertheless, on the margins of these major trends there has been a growing interest among archaeologists in the possibilities offered by the now enormous variety of Darwinian approaches to the study of past human societies and cultures. The mid-1990s seem to mark the coming of age of the archaeological application of such approaches. The past year has seen the publication of *Evolutionary Archaeology*, edited by Patrice Teltser; *The Archaeology of Human Ancestry: Power, Sex and Tradition*, edited by James Steele and myself; and now *Darwinian Archaeologies*, each demonstrating in its different way the potential these ideas have to offer.

As Herb Maschner and Steve Mithen make clear in their introduction, there are in fact some close parallels between Darwinian and post-processual approaches, ironically in view of the "political incorrectness" which remains (erroneously) associated with the former. Both emphasize individual intentions and decisionmaking; both accord social strategies as much importance as subsistence; both recognize the role of the contingencies of history and the importance of "heritage constraint." Even the poststructuralist emphasis on humans as caught in a web of signifiers beyond their control is mirrored by the idea of human minds as brains parasitized by populations of memes!

The political correctness issue though is not one that can be avoided and this seems as good a place as any to address it. It centers on the claim that the importation of Darwinian ideas into the human sciences is a form of ideology that seeks to justify the claim that biology is destiny, in support of a political agenda that is racist and sexist and seeks to roll back liberal legislation. Its rise in the last 20 years is explicable as a side effect of the rise of the right. The vast majority of those interested in the application of Darwinian ideas to anthropology and archaeology certainly do not see it this way, so what are the issues?

Perhaps the first one to be addressed should be the suggestion that Darwinian ideas may be interesting and relevant to the study of human societies but their potential political implications are too dangerous to be countenanced, so any discussion of them should be suppressed by some sort of socially responsible voluntary embargo. It seems to me that such a course of action is unlikely to achieve its desired aim and would in any event be an abdication of academic responsibility; once this is given up, academics lose their legitimacy, as their political opponents are only too quick to point out. In the case of evolution, the complexity of the political issues that arise, especially in North America, is very apparent, since many of those who want to keep Darwin out of anthropology would certainly want to defend a Darwinian account of the history of life in general against a creationist one.

But are the critics correct anyway in suggesting that Darwinian concepts have the socially and politically deleterious implications claimed? The answer provided by the papers in this volume is certainly "no." This is not to say that the ideas presented here must be correct. The argument is that they are interesting and potentially productive and should not be rejected out of hand on the basis of the erroneous belief that they have unpalatable implications. Let us look briefly at the main subareas of the Darwinian archaeology field identified by the contributors to this volume.

Cultural selectionism is an area of investigation that is following through the implications of analyzing the transmission of culture in similar terms to those that have been developed for understanding the transmission of genes. It makes few if any assumptions about the importance of biological reproductive success as an important goal of human social behavior. It is this latter topic that is the main focus of what may be called human sociobiology, which in fact figures little in this volume but has been the main focus of controversy. Those who pursue these studies are skeptical of the importance of culture to human societies, arguing that humans, like other creatures, behave in ways that are best explicable if one makes the assumption that their actions are oriented toward survival and reproductive success; local cultural goals are so closely correlated with the requirements of reproductive success that they can safely be left out of consideration. That individual human beings have some degree of innate propensity in such a direction seems to me at the very least not implausible, but it is certainly not the same as saying that particular social arrangements are thereby genetically determined. What does follow from assuming the existence of such propensities is that people will evaluate the costs and benefits of particular courses of action in the situations in which they find themselves in the light of them, and that, for example, the best course of action for a man in a particular context won't necessarily be the same as for a woman. But contexts change and balances of costs and benefits change with them: courses of action that may have been the best option in the past will not necessarily be so in the future. To suggest that the past legislates for the future in such situations is rather like saying that there is some rule that a particular population of plants or animals may not adapt to changing conditions.

In fact, probably the most widely held position—at least among archaeologists who are interested in Darwinian ideas—is that dual inheritance models are the most appropriate. These in effect argue that culture cannot be left out of account because it represents an independent transmission mechanism that leads to different outcomes from those predicted on the basis of reproductive success alone. On the other hand, built-in human propensities and their influence on the calculus of costs and benefits during individual decision making cannot be neglected either. Such an approach, which does not make any advance commitment as to what are the significant factors operating in any particular case but tries to leave them open to empirical determination, seems to me to be not too ideologically loaded.

Finally, we can turn to the remaining topic represented in this volume: Darwinian approaches to the evolution of mind. Clearly there are many different views on the extent to which the structure and content of the mind arise from the operation of Darwinian selection on the mental operations of previous generations, but here again, once we acknowledge the, at least to me, incontrovertible fact that the human brain is not a tabula rasa waiting to be imprinted with the structure of its local cultural context, then the issues become empirical ones that cannot be decided by ideological fiat in advance.

As Bettinger and Richerson point out in their overview, probably the most significant thing about the neo-Darwinian approach in biology is that it has led to an enormously significant and productive research program. Whether the same will turn out to be the case in archaeology remains to be seen, but the papers in this volume undoubtedly raise some of the key problems, such as Fletcher's interesting discussion of the nature of "memes." Furthermore, the papers also indicate the major conceptual retooling that will certainly be required at all levels of the discipline from data description upwards. Thus, archaeologists have always thought it part of their duty to define types, but such essentialist typological approaches mean that the intrapopulation variation that is central to any explanation based on selection does not even get described, never mind analyzed.

FOREWORD

This is only one of the stimulating issues that the contributions to this volume raise. If they convey something of the intimidating nature of the challenges posed by the creation of Darwinian archaeologies, they also get across a great deal of the excitement.

Stephen Shennan
Institute of Archeology
University College London
London, England

Preface

Darwinian theory has made a substantial impact on anthropology. It is either seen as the means by which anthropology might transcend its historical roots and become a force in the scientific study of humanity, or it is treated as the proverbial "straw man" and attacked out of hand, often by writers who have no understanding of Darwinian mechanisms. The papers in this volume elucidate some of the more salient points in taking a Darwinian approach while presenting the myriad of possibilities for applying Darwinian theory to archaeological problems. Thus, the papers included herein are Darwinian in nature but disparate in content—quite characteristic of an evolving field.

One of the major critiques of Darwinian approaches to archaeology has been the overabundance of rhetoric and the lack of application to actual archaeological data or problems of archaeological importance. This volume meets this challenge by including both theoretical discussions of Darwinian theory and a variety of papers that specifically address archaeological data and problems. These applications range from the origins of hereditary status differences to understanding stylistic variability to the nature of bifacial assemblages.

Some of the major themes addressed in this book include papers I have grouped under the heading of cultural selectionism—the notion that aspects of the human tool kit can be treated as extensions of the human phenotype. There are also a number of papers that deal with various forms of dual inheritance—the interplay between genetic evolution and cultural evolution. The rapidly evolving field of evolutionary psychology is also represented. Not included in this volume are specific discussions of evolutionary ecology or optimal foraging theory. Although these types of models are implied in many cases, they have been discussed in great detail elsewhere.

Throughout this volume I hope that the reader will find new ideas and enlightenment as we attempt to rejuvenate the science of prehistoric human behavior.

This volume has been a long time in preparation. Steve Mithen and I began planning the volume in 1990 and many of the papers were first drafted in 1991. I approached Mithen with the idea after reading his 1989 article "Evolutionary Theory and Post-Processual Archaeology" (*Antiquity* 63:483–494), which had put forward many of the same ideas I was formulating in my dissertation on Northwest Coast social inequality and warfare. This volume was set on the back burner several times, while both Steve (once) and I (twice) changed universities and academic positions. Fortunately, the authors showed great patience and understanding throughout the innumerable delays. Although Steve was unable to continue as an editor because of other commitments, the book slowly came together.

I would like to thank Steve Mithen for all his initial work on the book and all of the contributors for their patience and words of encouragement. I owe much to members of the faculty at the University of California at Santa Barbara for introducing me to evolutionary theory. These include Michael Jochim, Napoleon Chagnon, Donald Symons, Donald Brown, and John Tooby. I would like to thank series editor Michael Jochim and Plenum Executive Editor Eliot Werner for much urging and support. Caroline Funk worked many hours on the bibliographies and Lisa Frink did an excellent job of constructing the Index. I could not have completed this volume without their assistance. I would like to express my deepest appreciation to Bob Bettinger, whose enthusiasm and interest made this volume viable.

Finally, I would like to note the passing of a friend, Ben Cullen, December 29, 1995. We will miss him.

Herbert D. G. Maschner

Contents

I • INTRODUCTION

Chapter 1 • Darwinian Archaeologies: An Introductory Essay 3

Herbert D. G. Maschner and Steven Mithen

II • CULTURAL AND BEHAVIORAL SELECTION

Chapter 2 • The Historical Development of an Evolutionary Archaeology: A Selectionist Approach 17

Michael J. O'Brien

The Search for a Scientific Archaeology 19
Evolutionary Theory 23
Discussion 26
References 28

Chapter 3 • Explaining the Change from Biface to Flake Technology: A Selectionist Application 33

Alysia L. Abbott, Robert D. Leonard, and George T. Jones

Introduction 33
The Structure of Selectionist Thought 34
The Transition from Biface to Flake Technology: A Processual
    Explanation 35
A Selectionist Perspective on Technological Change 37

Conclusions 39
References 40

Chapter 4 • Cultural Virus Theory and the Eusocial Pottery Assemblage 43

Ben R. S. Cullen

Introduction 43
Science, Religion, and Cultural Virus Theory 45
Artifacts as Viral Phenomena 47
The Eusocial Pottery Assemblage 51
Conclusion 56
References 57

Chapter 5 • Organized Dissonance: Multiple Code Structures in the Replication of Human Culture 61

Roland Fletcher

Introduction 61
Cultural Messages 64
Multichannel Signal Replication 65
Basic Premises 68
Theoretical Framework 69
The Elementary Structure and Operation of Cultural Replication ... 71
The Locus of the Elementary Code Recipe 78
Operational Consequences 79
Conclusions 83
References 84

III • PATHS TO REVISIONISM IN CULTURAL–BEHAVIORAL SELECTION: INDIVIDUALS AND DUAL INHERITANCE

Chapter 6 • Kin Selection and the Origins of Hereditary Social Inequality: A Case Study from the Northern Northwest Coast 89

Herbert D.G. Maschner and John Q. Patton

Introduction 89

Culture as a Unit of Analysis	90
Kin Selection and Hereditary Social Inequality	92
A Northwest Coast Example	96
Discussion	100
Conclusion	101
References	102

Chapter 7 • Archaeology, Style, and the Theory of Coevolution 109

Kenneth M. Ames

Introduction	109
Dual Inheritance	110
Cultural Selection, Choice, and Imposition	115
Style	118
Northwest Coast Art	121
Summary and Conclusions	127
References	129

Chapter 8 • Style, Function, and Cultural Evolutionary Processes .. 133

Robert L. Bettinger, Robert Boyd, and Peter J. Richerson

Introduction	133
Style and Function in Neo-Evolutionary Perspective	135
A Critique of the Selectionist Program in Modern Archaeology	140
Processes of Stylistic Evolution	143
Conclusion	158
References	160

Chapter 9 • In Search of the Watchmaker: Attribution of Agency in Natural and Cultural Selection 165

Paul Graves-Brown

Introduction	165
The Verb "to Select"	166
The Attribution of Agency	167
Two Kinds of Struggle	169
Natural Selection: Cultural Selection	170
Natural Selection for Choice	171
Genesis: Epigenesis	172
Problems, Choices, Solutions	172

Dialectic with the Past: A Short History of Time 175
Conclusion: Purposes Divine and Profane 178
References 179

IV • COGNITION AND THE EVOLUTION OF MENTAL ADAPTATIONS

Chapter 10 • Weak Modularity and the Evolution of Human Social Behavior 185

James Steele

Introduction 185
Language and Cooperation as Weakly Modularized Abilities 186
Human Cognitive Evolution and the "Normal Social Environment" 189
Conclusions 193
References 194

Chapter 11 • The Origin of Art: Natural Signs, Mental Modularity, and Visual Symbolism 197

Steven Mithen

Introduction 197
The First Art 198
Symbols and the Capacity for Visual Symbolism 199
The Evolution of Visual Symbolism 200
Mark Making, Classification, and Communication 202
The Attribution of Meaning 203
ISVCs and Hunter-Gatherer Foraging 204
Modularity, Accessibility, and Hierarchization in Cognitive Evolution 207
Conclusion: Accessibility and the Middle/Upper Paleolithic Transition 212
References 213

V • OVERVIEW

Chapter 12 • The State of Evolutionary Archaeology: Evolutionary Correctness, or the Search for the Common Ground 221

Robert L. Bettinger and Peter J. Richerson

Paradigms Require Paradigmatic Case Studies 223

CONTENTS

Evolutionary Ecology Is Evolutionary Research 223
Models *Are* Just-So Stories 225
Intent Is an Appropriate Subject of Study and Component of Theory 226
The Emergent Properties of Groups Should Not Be Ignored 228
Analogues between Evolutionary Process in Natural and Cultural
 Systems Are Just Analogues 229
Do the Math 229
References 231

Index **233**

Darwinian Archaeologies

PART I

INTRODUCTION

Chapter 1

Darwinian Archaeologies
An Introductory Essay

HERBERT D. G. MASCHNER AND STEVEN MITHEN

The significance of evolutionary theory for archaeology is fiercely debated: one does not entitle a theory or interpretation, let alone a book, as "Darwinian" without inviting controversy. For many archaeologists the word *Darwinism* implies a denial of our humanity and free will and is thought to support a pernicious political agenda by legitimizing selfish individualism. For others, the word invokes quite different ideas: it suggests an attempt to view ourselves as part of, rather than separate from, the natural world with the many positive political, social, and economic implications that would follow. Others remain unmoved. They simply feel that any reference to biological evolution is irrelevant to the explanation of human behavior and culture change. In short, there is much controversy and little agreement.

So let us start this book with a statement with which we hope all can agree: the human species is a product of biological evolution. We were not created by divine intervention, but evolved by the same processes as other species. Of these natural selection is likely to have played a major role. We doubt if many archaeologists would question the above assertions—there is little dispute about this "evolutionary fact" (Ruse 1986:1). It is when archaeologists are asked about the relevance of this evolutionary fact for their discipline that one

---

HERBERT D. G. MASCHNER • Department of Anthropology, University of Wisconsin, 1180 Observatory Drive, 5240 Social Science, Madison, Wisconsin 53706. **STEVEN MITHEN** • Department of Archaeology, University of Reading, Whiteknights, P.O. Box 218, Reading RG6 2AA, United Kingdom.

finds the immense range of responses and strong feelings to which we referred above. These cover the whole spectrum from total irrelevance to the belief that this evolutionary fact should condition the manner in which we pursue our discipline. We fall toward the latter end of this spectrum.

We believe that a full understanding of human behavior and culture change requires an explicit reference to our biological constitution and the processes by which it evolved. In itself, this is not a particularly grand statement. Reference to evolutionary theory is clearly insufficient as a *complete* explanation for any form of behavior or event. In any one instance there are many processes and phenomena to which reference must be made before we acquire anything approaching a full understanding. Indeed, there are likely to be many aspects of human behavior for which reference to evolutionary theory may be relatively insignificant although it remains unclear to us at present which areas these might be. But we start this book with the simple assumption that for many aspects of human behavior reference to evolutionary theory is essential. Until such reference is made, explanation will remain incomplete. This belief constitutes the first of two rationales for this volume.

In this regard we have brought together a series of papers which explore the nature of a Darwinian archaeology through a combination of theoretical argument and case studies. One of our aims in bringing these papers together is to demonstrate the breadth of subject matter to which Darwinian approaches are relevant. Consequently, within this volume the archaeological references stretch from the earliest Paleolithic to the Historic period as arguments are made that a Darwinian perspective will illuminate, among other things, the development of prehistoric stone technology, the origin of art, and the emergence of social inequality.

Yet these studies were brought together for a purpose beyond that of blowing a Darwinian trumpet. A second, and more important, rationale for the volume is to explore the most appropriate form that a Darwinian archaeology should take. If we were to take all archaeologists who agreed that the fact of human evolution is indeed significant for human behavior and culture change, and were to ask them just *how* it is significant, a second spectrum of replies would be found. This spectrum would be just as wide and as diverse as the first. It is this issue, the *character* of a Darwinian archaeology, that lies at the heart of this volume, rather than the issue of whether or not a Darwinian archaeology (whatever it may look like) is required.

Consequently, these papers were collected not only to illustrate, but also to explore the diversity of views concerning the current use of evolutionary theory in archaeology. We hope the volume facilitates the comparison and evaluation of these approaches and fosters the development of a greater consensus in the manner in which evolutionary theory can be drawn on to aid our understanding of the past. At present, such consensus appears remote. We ourselves are unsure as to the most appropriate direction in which such studies should proceed and the criteria that should be used for evaluation. Consequently, this volume is entitled *Darwinian Archaeologies*, to stress that there is no single Darwinian approach that is sufficiently widely accepted to be awarded the title of *the* Darwinian archaeology.

We feel that these papers illustrate the breadth of approaches that currently fall under the Darwinian umbrella. When the idea for this book was conceived, in 1991, we were both impressed, and slightly bewildered, by the number of alternative Darwinian approaches that were appearing in the literature. Theories and methods of a remarkably different character were all claiming to be a Darwinian, some to be *the* Darwinian, approach to culture change. We recognized the need to bring these studies together to provide a firm base for comparison and evaluation.

As little as a decade ago this would have been unnecessary. During the 1970s and early 1980s an ecological/functionalist approach developed in archaeology, particularly with regard to hunter-gatherers among American scholars. This lay at the heart of processual archaeology and claimed the Darwinian mantle. This archaeology frequently invoked concepts of selection and adaptation and sought theoretical justification by reference to evolutionary theory. Much of this is summed up in White's (1949) classic statement, which was championed by Binford (1972): culture is humanity's "extrasomatic means of adaptation." The reference to adaptation was critical in making an implicit reference to evolutionary theory and the natural sciences, in which the legitimation for the principles of processual archaeology was found. The Cambridge Paleoeconomy school (Jarman *et al.* 1982) had done much the same.

The notion of adaptation lay at the heart of this approach and was conceived as relating to the behavior of groups, with specific reference to the balance between people and their resources. Subsistence practices, technology, social organization, and even art were interpreted as functioning to "fit" the group to the availability and distribution of resources in the natural environment. In this approach, there was a limited role for human agency or intentionality. People were purely reactive in nature: "selection for change occurs when the system is unable to continue previously successful tactics in the face of changed conditions in the environment" (Binford 1983:203). While we question its Darwinian credentials, we have no desire to dismiss the invaluable contribution such work has made to the development of archaeological thought. For instance, the papers in the classic volume edited by Bailey (1983) exemplify this group adaptationist approach in relation to hunter-gatherers. They demonstrate the very substantial contribution that it has made to our study of prehistoric hunter-gatherers.

A path of development can be traced from qualitative descriptions of past behavior as adaptations to more formal quantitative studies (e.g., Jochim 1976) which found their natural home in optimality theory (e.g., Winterhalder and

Smith 1981) in which reference to evolutionary theory became explicit. This became most widespread in the study of prehistoric foraging either by the use of concepts and terminology taken directly out of the optimal foraging literature of behavioral ecology or by the attempt to build quantitative models for past behavior. Torrence (1986) tried to develop a similar approach to technology. The majority of studies maintained the focus on group behavior (e.g., Belovsky 1987), even though the contradiction between this and the theoretical justification for such modeling (relying on an individualistic perspective, see Stephens and Krebs 1986) became very apparent.

This type of Darwinian Archaeology constitutes what is still thought of as *the* Darwinian approach in archaeology: a concern with group behavior, with subsistence, and which is, in essence, functionalist. Indeed Darwinian archaeology is often taken to be synonymous with functionalism (e.g., Shanks and Tilley 1987).

The group adaptationist approach was joined in the 1980s by another Darwinian archaeology that explicitly advertised itself as the application of evolutionary theory in archaeology: cultural selectionism.[1] Dunnell (1980) provided a seminal paper for this approach which has received extensive development and application to culture change (e.g., Rindos 1984; Leonard and Jones 1987; Dunnell 1989; Leonard 1989; O'Brien and Holland 1990, 1992). The approaches which adopt a cultural selectionist label are diverse but perhaps the central and common feature is the willingness to treat artifacts as part of the human phenotype, even though they are physically detached from the human body. Following from this, it is argued that the "fitness" of an artifact can be measured by its replication and spread through space and time and consequently a selectionist terminology becomes an appropriate framework for interpreting the archaeological record. Rindos (1984) provides a particularly detailed discussion of cultural selectionism and its relationship to biological evolution and uses this to interpret the origins of agriculture.

Cultural selectionism, in its many guises, remains one of the most prevalent forms of Darwinian archaeology currently on the agenda. Consequently, four papers are included in this volume that can be characterized as a cultural selectionist archaeology. The first is a historical treatise on the place of evolutionary approaches in modern scientific archaeology by Michael J. O'Brien. He argues that one of the primary failings of 1960s archaeology was a lack of adherence to the evolutionary principles being advocated. A second selectionist paper is principally a case study by Alysia L. Abbot, Robert D. Leonard and George T. Jones of the transition from a biface to flake technology in ancient North America. They contrast a cultural selectionist against a processual explanation for this transition arguing that the former has greater explanatory pow-

---

[1] This is not a term generally used by the proponents of the approach.

er. The next two papers are primarily theoretical. Ben Cullen proposes an extreme variant of cultural selectionism which he terms *cultural virus theory*. This is a radical reinterpretation of cultural selectionism in which the elements of human culture, especially ideas, are described as "parasitic" and are replicated and transformed by human minds. Following this is a paper by Roland Fletcher who, building on the type of ideas presented by Cullen, focuses on the processes by which cultural units are replicated. He suggests that the human cultural repertoire should be thought of as composed of "messages" of different types, such as words, actions, and material things. Each type of message has its own unique characteristics, such as its rate of transmission, but together they can be thought of as forming an "assemblage of cultural features" that is replicated and transformed.

Since the later 1980s the group adaptationist and cultural selectionist approaches have been joined by a number of other Darwinian archaeologies. Three themes are apparent: attempts to focus on the individual; a concern with the interaction between biological and cultural modes of inheritance; a cognitive approach drawing on an evolutionary approach to human psychology.

Those attempting to focus on the individual rather than the group (e.g., Mithen 1990; Maschner 1992) claim to be attempting to conform to the biological definition of the term *adaptation* which they suggest requires reference to individual behavior rather than groups. Following evolutionary biologists, the whole notion of group adaptation is rejected unless it is conceived of being no more than the summation of individual adaptations and, as such, would have little analytical value in itself (e.g., Jochim 1981:16; Mithen 1989, 1993). This marks an attempt to move away from cultural ecology to a more explicitly evolutionary ecology for past human behavior. In such work the "stable until pushed" premise of processual archaeology is replaced by a view that societies are always in a state of readjustment, experimentation, and change because certain individuals within the society will be attempting to manipulate it to his/her own ends. In this volume Maschner and Patton extend the arguments for an individualistic perspective by presenting a Darwinian explanation for the origins of hereditary social inequality, with specific reference to the Northwest coast of the United States. This centers on the role of individuals striving for social prestige and power by the use of coercive power.

Those studies which focus on the interaction between social learning, cultural transmission, and biological evolution find their roots in the mathematical studies of Cavalli-Sforza and Feldman (1981) and Boyd and Richerson (1985). Durham (1991) developed a more qualitative approach and termed this dual inheritance theory for its essence lies in understanding the dynamic interplay between the biological and cultural modes of inheritance. Kenneth Ames's paper in this volume provides a comprehensive summary and discussion of dual inheritance theory and contrasts it with cultural selectionism. These approaches are compared in terms of how they interpret style in the archaeological record, with Ames arguing that the dual inheritance approach is far more productive than the pure cultural selectionism of Dunnell (1978) and O'Brien and Holland (1992). Building on their previous works, Boyd and Richerson (1985) and Bettinger (1991) join together to present a joint view of their view of an evolutionary archaeology. They argue that one of the areas where a Darwinian approach may be most useful is in the study of the relationship between style and function. They further state that it is only a matter of time before Darwinian theory will produce an understanding of culture evolution as powerful as the one it now provides to genetic evolution.

A third new strand of Darwinian archaeologies are those studies which involve an explicit reference to the human mind as a product of biological evolution. These find their roots in the growing interaction between biological and psychological approaches to behavior. Hinde (1987) coined the term *psychological propensities* to refer to features common to all human minds. Others have referred to *Darwinian algorithms*, and sought to define a new discipline of *evolutionary psychology* (Cosmides and Tooby 1987; Barkow et al., 1992). The fundamental premise in this work is that the structure of the human mind, with regard to these propensities or algorithms, evolved in a Pleistocene environment when our human ancestors were faced with a radically different range of problems than we face today in our modern surroundings. Consequently, for many types of modern behavior, there is very limited expectation that it will be "adaptive" in any biological sense. Much of our cultural behavior, many aspects of which are patently nonfunctional and may actually be maladaptive, derives from the crunch between a Pleistocene psychology and an urban sociology (Maschner 1992, 1996).

Two papers in this volume can be classed into this domain of a cognitive Darwinian archaeology. James Steele begins with the notion that a Darwinian archaeology must rely on the assumption of a set of universal predispositions and explores the evidence for their existence and evolution in the fossil record. He explores whether manual dexterity, language, and social interaction might be based on such predispositions and discusses the implications for a Darwinian archaeology. Mithen is also concerned with the timing of the evolution of particular cognitive structures. In his case he is not concerned so much with the evolution of discrete cognitive structures or psychological propensities, but rather with what he sees as the increasing connections between such structures in recent cognitive evolution. He integrates this with an evolutionary ecological approach to symbolism suggesting that this cognitive Darwinian approach has a major contribution for our understanding of the origin of art.

All three of these new strands of Darwinian archaeologies, the individualistic, dual inheritance, and evolutionary psychological approaches, share many similarities and overlap in their use of specific concepts. We suspect that, along with aspects of cultural selectionism, they constitute the critical elements for a new synthesis for Darwinian archaeology. Yet that synthesis remains to be built. At present there remains major disagreements between these Darwinian archaeologies, and those of group adaptation. These become readily apparent in this book and several authors explicitly criticize or challenge other Darwinian archaeologies. Let us emphasize three major areas of dispute.

The first is simply what constitutes the appropriate unit for analysis. Three options are found within Darwinian archaeologies. First, there are the cultural elements themselves, as in the work of cultural selectionists. This treats material culture, or in Cullen's case ideas, as part of the human phenotype and as a result considers them directly subject to selective pressure. This is a logical extension of the arguments put forward by Dawkins (1982) in his book *The Extended Phenotype*. One of the appeals of this approach is that it makes a direct link to the traditional concerns of archaeologists in the form of artifact typologies and culture histories. But to many, an archaeology without people as its central focus is inherently flawed.

When people are placed in this role, there is disagreement as to the relative merits of an individual or a group approach. There has been considerable discussion within the evolutionary ecological literature concerning the unit of selection and substantial consensus is found that it is the individual, not the group (Williams 1966; Krebs and Davies 1987). Wynne-Edwards (1962, 1986) is almost the only dissenting voice and group selection and group adaptation appear only possible in exceptional circumstances. This conclusion runs directly at loggerheads with the dominance of group adaptationist approach in archaeology. As noted above, some archaeologists represented in this volume, have tried to shift the archaeological studies to be more agreeable with evolutionary theory by adopting an individualistic perspective. But they meet stiff opposition from those who believe that group adaptation is the most appropriate stance for archaeology (e.g., Clark 1992).

In spite of the greater theoretical integrity of an individualistic approach, three major arguments can be mustered in support of a group perspective. The first is simply the major contribution that group adaptation has made in archaeology, particularly that of prehistoric hunter-gatherers. If we were to judge group adaptation by its record, then it becomes a very persuasive Darwinian archaeology. Second, one can point to the daunting operational problems encountered by archaeologists when attempting to focus on individuals, whether or not they qualify this by referring to *generic* rather than *specific* individuals (e.g., Mithen 1993). It is simply impossible to monitor and track individuals in the archaeological record and an individualistic perspective may be running archaeology into an operational brick wall. Third, if it is acknowledged that group adaptation can only happen in special circumstances, then it might be argued that it is precisely those circumstances that make human sociality unique.

A fourth argument might also be put forward. To what extent do archaeologists need to adhere to the definition placed on terms such as *adaptation* by biologists? Can we not create and use our own definitions that are suitable for our own particular discipline? After all, the manner in which the adaptation is commonly used and understood in everyday speech is much closer to the archaeological than biological usage of the term. Is it not legitimate to create our own definitions of terms such as adaptations, as in the manner of Clark and Straus (1983:136)? This clearly touches on a much larger issue concerning the use of language.

We believe that both group and individual oriented approaches have a major role to play in Darwinian archaeology, though our own preference is with the latter. But whichever one chooses, there is a second major fault line between Darwinian archaeologies. This concerns the terminology in which we frame our evolutionary arguments and centers around the term *selection*.

For some a selectionist approach and terminology is central to a Darwinian archaeology. Indeed, O'Brien and Holland (1992) use the extent to which a trait is selected to define whether or not it should be considered as an adaptation. In these approaches, human intentionality can largely be ignored; this acts as only the source of variation and is directly analogous to the role of genetic mutation in biological evolution. According to O'Brien and Holland, human intentionality cannot be denied but is "trivial" for framing Darwinian explanations. There is a close similarity here to the manner in which processual archaeologists pursuing group adaptationist explanations dismissed the role of human agency in culture change.

Within other Darwinian archaeologies this language of selection and the dismissal of human intentionality is questioned. They note that, unlike mutation in biological evolution, human choice acts not only as the source of variation but also as the mechanism of selection. People do indeed choose how to act and their behavior is goal directed. Any analogy between human choice and natural selection is ill-founded. Some choices are conscious and result from careful weighing up of the alternatives on offer; others are unconscious and derive from the following of tradition. But to dismiss this intentionality as trivial and to force human actions into a selectionist framework in which we lose sight of people going about their everyday actions appears inappropriate. On the one hand, it risks presenting people as helpless automatons in the face of an overpowering mechanistic selective force; on the other, it is simply counterintuitive and does not accord with our own experience of what it means to be human. As a consequence, rather than adopt a terminological framework of selection, we need an explicit reference to human agency. To this end, reference might be made to the processes of decision making, problem solving, and learning. These are the proximate means by which human adaptations arise. Evolutionary ecologists studying nonhuman animals are also increasingly finding it necessary to refer to proximate causation (e.g., Dunbar 1984).

In this volume, Graves provides an in-depth discussion of the relationship between agency and selection in evolution. And he shows that our concern with the relationship between cultural and biological evolution has a long and distinguished pedigree for he discusses how Darwin himself struggled to resolve the awkward relationship between agency and selection. Graves takes many Darwinian approaches to task for failure to sufficiently integrate the individual into his/her social context. As he argues, an excessive reliance on free will and choice is as erroneous as overemphasis on blind natural selection.

These disagreements about the appropriate unit for study and terminology to adopt pervade this volume and demonstrate the substantial contrasts between different Darwinian archaeologies. We hope that by bringing these together we will facilitate the comparison and evaluation of these different approaches. Such work will no doubt be undertaken by those already sympathetic to the notion of a Darwinian archaeology. But we also want to address this volume to the "others"; to those who agreed to the fact of human evolution, but who believe that this is insignificant for understanding the manner in which humans behave and the process of culture change.

To this group of archaeologists we hope that this volume can clarify three major misconceptions about the use of evolutionary theory in archaeology. We hope that by doing this we can broaden the range of archaeologists sympathetic to the use of evolutionary theory, even though they themselves may wish to keep their distance.

The first misconception relates directly to the major theme of this book—that there is no single Darwinian archaeology. There are many different approaches on offer hosting a wide range of concepts and methods. As such, Darwinian archaeology is not a discrete, bounded approach that can be safely characterized, criticized, and forgotten, but one that must be central to the debate of what constitutes an appropriate archaeology. If one wishes to understand and discuss the role of theory in archaeology, one has no choice but to immerse oneself within the diversity of Darwinian archaeologies.

The second misconception is the equivalence between a Darwinian and a functional interpretation of behavior. The collapse of the link between a Darwinian and a functionalist interpretation of behavior is perhaps the most significant development in recent archaeological thought. Certainly if we are to explain how a behavior that helps an individual or group survive and reproduce (i.e., is functional in a biological sense), we would need to invoke some sort of Darwinian explanation. But this does not mean that whenever a Darwinian approach is presented, one is necessarily seeking to interpret behavior in a functional sense. Indeed it is one of the fundamental features of the cultural transmission and evolutionary psychology strands of Darwinian archaeology that these directly address, and help explain, nonfunctional behavior.

A third misconception is that a Darwinian archaeology necessarily carries the same philosophical baggage as the processual archaeology of the 1970s. This position is understandable. The invoking of evolutionary theory was a critical part of processual archaeology, along with the desire for particular forms of explanation involving hypothesis testing, functionalism, and, in certain early work, a call for "laws" of culture process. Yet there is no necessary link. Just as to adopt a Darwinian archaeology need not imply a functionalist interpretation of behavior, it need not imply any adherence to the forms of explanation and the particular archaeological concerns associated with processual archaeology.

Indeed, one of the most noticeable features of the more recent Darwinian archaeologies is that they share so many interests with the archaeologies proposed by the self-proclaimed postprocessualists (e.g., Hodder 1985). One major similarity is that both domains of thought have argued that greater attention ought to be paid to the individual when attempting to understand the nature of society. Neither feel that this is the sole analytical unit to adopt and recognize that the relationship between the individual and society is complex.

A second point of contact is with the emphasis on social strategies. The themes of power, domination, and exploitation have been important in postprocessual archaeology (e.g., Miller and Tilley 1984). These are similarly stressed within evolutionary theory in which as much, if not greater, stress is laid on the adaptation to the social environment (which often involves the adoption of social strategies to acquire power, wealth, and prestige) as to the physical environment in terms of exploiting natural resources (e.g., Betzig et al. 1988). This concern with social strategies is particularly represented in Maschner and Patton's discussion of the origin of social inequality.

A third point of contact can be found with regard to the recognition of the importance of unique historical contexts in which individual actions take place. Contingency can play a major role in Darwinian interpretations of culture change (e.g., Mithen 1989, 1990; O'Brien and Holland 1992). This is stressed by numerous authors in this volume, including Cullen and Fletcher, in terms of "heritage constraint," and Mithen and Graves-Brown.

In these regards, therefore, the emerging Darwinian archaeologies share many of the concepts and concerns with postprocessual archaeologies. Ironically, the latter have often partly sought their identity by rejecting the validity of an evolutionary perspective on human behavior. We can see that this rejection is based on a misunderstanding that Darwinian archaeology is necessarily associated with functionalism and the concerns and approaches of a simplified processual archaeology. As such, there has been a failure to recognize that many strands of Darwinian archaeology are concerned with the issues of power, cognition, and unique historical paths of cultural development. Consequently, we hope that this volume may help remove this narrow categorizing of evolutionary approaches and help engender a greater dialogue between archaeologists who find their intellectual roots in either the social or the natural sciences.

REFERENCES

Bailey, G.N. (ed.), 1983, *Hunter-Gatherer Economy in Prehistory*, Cambridge University Press, Cambridge.
Barkow, J., Cosmides, L., and Tooby, J. (eds.), 1992, *The Adapted Mind: Evolutionary Psychology and the Generation of Culture*, Oxford University Press, Oxford.
Belovsky, G., 1987, Hunter-Gatherer Foraging: A Linear Programming Approach, *Journal of Anthropological Archaeology* 6:29–76
Bettinger, R., 1991, *Hunter-Gatherers: An Evolutionary Approach*, Plenum Press, New York.
Betzig, L., Borgerhoff Mulder, M., and Turke, P., 1988, *Human Reproductive Behavior: A Darwinian Perspective*, Cambridge University Press, Cambridge.
Binford, L.R., 1972, *An Archaeological Perspective*, Seminar Press, New York.
Binford, L.R., 1983, *In Pursuit of the Past*, Thames and Hudson, London.
Boyd, R., and Richerson, P.J., 1985, *Culture and the Evolutionary Process*, University of Chicago Press, Chicago.
Cavalli-Sforza, L.L., and Feldman, M.W., 1981, *Cultural Transmission and Evolution: A Quantitative Approach*, Princeton University Press, Princeton.
Clark, G.A., 1992, A Comment on Mithen's Ecological Interpretation of Paleolithic Art, *Proceedings of the Prehistoric Society* 58:107–109.
Clark, G.A., and Straus, L., 1983, Late Pleistocene Hunter-Gatherer Adaptations in Cantabrian Spain, in: *Hunter-Gatherer Economy in Prehistory* (G.N. Bailey, ed.), Cambridge University Press, Cambridge, pp. 131–148.
Cosmides, L., and Tooby, J., 1987, From Evolution to Behavior: Evolutionary Psychology As the Missing Link, in: *The Latest on the Best; Essays on Evolution and Optimality* (J. Dupre, ed.), MIT Press, Cambridge, MA, pp. 277–306.
Dawkins, R., 1982, *The Extended Phenotype*, Oxford University Press, Oxford.
Dunbar, R., 1984, *Reproductive Decisions. An Economic Analysis of Gelada Baboon Social Strategies*, Princeton University Press, Princeton.
Dunnell, R., 1978, Style and Function: A Fundamental Dichotomy, *American Antiquity* 43:192–202.
Dunnell, R., 1980, Evolutionary Theory and Archaeology, in: *Advances in Archaeological Method and Theory*, Vol. 3, (M.B. Schiffer, ed.), Academic Press, New York, pp. 35–99.
Dunnell, R., 1989, Aspects of the Application of Evolutionary Thought in Archaeology, in: *Archaeological Thought in America* (C.C. Lamberg-Karlovsky, ed.), Cambridge University Press, Cambridge, pp. 35–49.
Durham, W.H., 1991, *Coevolution: Genes, Culture and Human Diversity*, Stanford University Press, Stanford.
Hinde, R., 1987, *Individuals, Relationships and Culture*, Cambridge University Press, Cambridge.
Hodder, I., 1985, Post-Processual Archaeology, in: *Advances in Archaeological Method and Theory*, Vol. 8 (M.B. Schiffer, ed.), Academic Press, New York, pp. 1–25.

Jarman, M.R., Bailey, G.N., and Jarman, H.N., 1982, *Early European Agriculture: Its Foundations and Development*, Cambridge University Press, Cambridge.
Jochim, M., 1976, *Hunter-Gatherer Settlement and Subsistence*, Academic Press, New York.
Jochim, M., 1981, *Strategies for Survival*, Academic Press, New York.
Krebs, J.R., and Davies, N.B., 1987, *An Introduction to Behavioral Ecology*, Blackwell, Oxford.
Leonard, B., 1989, Resource Specialization, Population Growth and Agricultural Production in the American Southwest, *American Antiquity* 54:491–503.
Leonard, B., and Jones, G.T., 1987, Elements of an Inclusive Evolutionary Model for Archaeology, *Journal of Anthropological Archaeology* 6:199–219.
Maschner, H.D.G., 1992, The Origins of Hunter and Gatherer Sedentism and Political Complexity: A Case Study from the Northern Northwest Coast, Unpublished Ph.D. dissertation, University of California, Santa Barbara.
Maschner, H.D.G. (1996). The Politics of Settlement Choice on the Northwest Coast: Cognition, GIS, and Coastal Landscapes, in: *Anthropology, Space, and Geographic Information Systems* (M. Aldenderfer and H. Maschner, eds.), Oxford University Press, Oxford, pp. 175–189.
Miller, D., and Tilley, C., 1984, Ideology, Power and Long Term Social Change, in: *Ideology, Power and Prehistory* (D.Miller and C.Tilley eds.), Cambridge University Press, Cambridge, pp. 147–152.
Mithen, S., 1989, Evolutionary Theory and Post-processual Archaeology, *Antiquity* 63:483–494.
Mithen, S., 1990, *Thoughtful Foragers: A Study of Prehistoric Decision Making*, Cambridge University Press, Cambridge.
Mithen, S., 1993, Individuals, Groups and the Paleolithic Record: A Reply to Clark, *Proceedings of the Prehistoric Society* 59:393–398.
O'Brien, M., and Holland, T., 1990, Variation, Natural Selection and the Archaeological Record. in: *Advances in Archaeological Method and Theory*, Vol. 2, (M.B. Schiffer, ed.), University of Arizona Press, Tucson, pp. 31–79.
O'Brien, M., and Holland, T., 1992, The Role of Adaptation in Archaeological Explanation, *American Antiquity* 57:36–59.
Rindos, D., 1984, *The Origins of Agriculture: An Evolutionary Perspective*, Academic Press, New York.
Ruse, M., 1986, *Taking Darwin Seriously*, Blackwell, Oxford.
Shanks, M., and Tilley, C., 1987, *Reconstructing Archaeology*, Cambridge University Press, Cambridge.
Stephens, D., and Krebs, J., 1986, *Foraging Theory*, Princeton University Press, Princeton.
Torrence, R., 1986, *Time, Energy and Stone Tools*, Cambridge University Press, Cambridge.
White, L., 1949, *The Science of Culture*, Grove Press, New York.
Williams, G.C., 1966, *Adaptation and Natural Selection*, Princeton University Press, Princeton.
Winterhalder, B., and Smith, E.A., 1981, *Hunter-Gatherer Foraging Strategies*, University of Chicago Press, Chicago.
Wynne-Edwards, V.C., 1962, *Animal Dispersion in Relation to Social Behavior*, Oliver & Boyd, Edinburgh.
Wynne-Edwards, V.C., 1986, *Evolution through Group Selection*, Blackwell, Oxford.

PART II

CULTURAL AND BEHAVIORAL SELECTION

Cultural selection is not a term generally used by most of the authors in this section. Yet, the notion that items of material culture can be treated as extensions of the human phenotype, and the implication that successful variants of technology are a result of selective pressures on decision making, leads to that term. Cultural selection developed from the need to address technological change and variation in a scientific manner. It further resulted from a desire to investigate change from a Darwinian perspective. Although, in the next paper, O'Brien argues that we do not need to call this approach *cultural selection*, the term does distinguish these papers from the sections that follow.

Originally developed by Dunnell, this approach is generally considered the standard approach to Darwinian archaeology. While approaches that embrace a cultural selectionist label are diverse, their central theme is that the "fitness" of an artifact is measured by its replication and spread through time and across the prehistoric landscape. Thus, a selectionist nomenclature is appropriate for addressing culture change. Cultural selectionism, in its many forms, remains one of the most prevalent forms of Darwinian archaeology currently on the agenda and has been used to address issues as broad as the origins of the state and foraging ecology of hunters and gatherers.

Chapter 2

The Historical Development of an Evolutionary Archaeology
A Selectionist Approach

MICHAEL J. O'BRIEN

Increasingly in the last several years there has been a growing number of archaeologists who are beginning to take note of the fact that Darwinian evolution offers a powerful means of explaining variation in the material record. The approach has been variously termed *evolutionary*, or *selectionist*, *archaeology*, and though it is still in a formative stage, there are clear signs of future growth and development. Although Darwinian evolutionary archaeology has not enjoyed the meteoric rise seen in the overnight sensation of the 1960s, processual archaeology, there are now in preparation or in press several edited books on the subject (e.g., Teltser 1995; O'Brien 1996), as well as numerous evolutionarily focused articles in leading archaeological journal (e.g., Dunnell 1978a, 1980; Leonard and Jones 1987; Rindos 1989; O'Brien and Holland 1990, 1992; Neff 1992; O'Brien *et al.* 1994) and monographs (e.g., Feathers 1989; Braun 1990; Dunnell 1992, 1995; O'Brien and Holland 1995a,b).

Despite the attention that scientific evolutionism in receiving from archaeologists, unless the discipline understands the basic tenets of the approach and is convinced of its power in explaining variation in the archaeological record, there is no reason to suspect that it will be widely accepted. The field of archaeology is a veritable graveyard of paradigms that have waxed and waned

---

MICHAEL J. O'BRIEN • Department of Anthropology, University of Missouri, 200 Swallow Hall, Columbia, Missouri 65211.

over the years, and, without considerable effort to show that evolutionary theory is the *only* means available for actually explaining the archaeological record as opposed to simply interpreting it, evolutionary archaeology will simply be another in the long line of casualties. It, unlike many other approaches that have been proposed in the discipline, needs full participation by researchers. This is because no other approach requires such a massive amount of meticulously generated data. Its success—measured in terms of its own performance in successfully explaining the archaeological record—depends on technological and functional data over which there exists tight temporal control. Unfortunately, most currently available data are not useful in addressing evolutionary questions because they were not generated for that purpose.

Data generated to examine culture-historical issues (including chronological ordering), while often suited to that purpose, cannot legitimately be extended to examination of issues of function. And it appears obvious that it is the functional aspects of the archaeological record that are most readily incorporated into a scientific evolutionary framework (Haag 1959; Dunnell 1978a,b; O'Brien and Holland 1990, 1992; O'Brien *et al.* 1994). The fact that currently available archaeological data are, for the most part, inappropriate for inclusion in a scientific archaeology might sound counterintuitive to some archaeologists—a situation that appears to have arisen from confusion over fundamental differences between typological thinking, with its emphasis on *transformation*, and population thinking, with its emphasis on *variation* and *replacement*. This issue is as pertinent to archaeology (Dunnell 1980, 1982, 1985, 1986, 1988; O'Brien and Holland 1990, 1992; O'Brien *et al.* 1994) as it is to biology (Dobzhansky 1951; Mayr 1963, 1973, 1976, 1977; 1987; Ghiselin 1966, 1974, 1981; Sober 1980, 1984).

Clearly, as Dunnell has pointed out *ad nauseam* (e.g., Dunnell 1989a), evolution is a materialist strategy that has its roots in population thinking. Equally clearly, archaeologists still do not understand the ramifications of this statement. We still speak of *types*, for example, not as theoretical units but rather as empirical units, i.e., as "real" things. How can this be, if we are seriously interested in incorporating scientific evolution in archaeology? Could it be that we do not understand the difference between theoretical units and empirical units? Are we missing the distinction between *essentialism*—which by its very name signals an interest in the "essential" qualities that something possesses—and *materialism*? At another level, are we viewing science as a monolithic entity, not realizing that there is a world of difference between physical science and the things in which it is interested and life science and its fields of interests?

Perhaps this is a good place to review briefly what evolutionary archaeology is and what it is not. Collections of essays such as this one play an important role in furthering a general understanding of the scientific evolutionary approach, if for no other reason than they cause its proponents to rethink their position in terms of clarity and logic. I say this is a good place for reviewing the tenets of scientific evolution precisely because of the growth in its visibility. There is, however, a danger in its increased visibility. History bears out that archaeologists, for a variety of reasons, have been quick to jump on bandwagons without the slightest notion why they are doing it and certainly without the background necessary to understand the nuances of the approaches they begin advocating. Lessons learned from archaeology conducted in the 1960s and 1970s are informative here and perhaps shed light on the question of why scientific evolution has only recently begun to be incorporated in archaeology. Perhaps more importantly, though space precludes anything more than brief mention, we can also learn from the lessons of biologists in the 1930s and 1940s as they grappled with evolutionary issues.

THE SEARCH FOR A SCIENTIFIC ARCHAEOLOGY

Archaeologists of the 1960s and 1970s might have understood what Lewis Binford (e.g., 1962, 1968) and others meant by the term *culture process* (e.g., various papers in S. R. Binford and L. R. Binford 1968), but many of them certainly did not understand exactly what it meant to be a processual archaeologist. In fact, only now is it becoming clear what archaeologists really were accepting when they called themselves processualists. Beyond question, the discipline was becoming increasingly bored with a singular focus on issues such as time and space—a movement that can be traced back at least 25 years before Binford (1962) wrote "Archaeology as Anthropology" (e.g., Steward and Setzler 1938; Steward 1942; Bennett 1943; Taylor 1948; Caldwell 1958; Willey and Phillips 1958; Willey 1962). Archaeologists, at least some of them, were concerned that their discipline had, in its emphasis on time-and-space systematics, overlooked culture—that nebulous concept that makes us human and which had, by the middle of the 20th century, become the unifying principle of anthropology. In fact, Binford (1962:217) began his essay "Archaeology as Anthropology" by praising Willey and Phillips's (1958:2) famous quote "American archaeology is anthropology or it is nothing" and then proceeding to state that "the purpose of this discussion is to evaluate the role which the archaeological discipline is playing in furthering the aims of anthropology and to offer certain suggestions as to how we, as archaeologists, may profitably shoulder more responsibility for furthering the aims of our field."

Furthering the aims of "our" field, indeed. Through the efforts of Binford and others, especially Kent Flannery (e.g, 1968a,b, 1972; Flannery and Coe 1968), archaeologists soon learned that culture could indeed be added back to the equation through such things as ecology and general systems theory, and they began not only to steep themselves in the principles of those disciplines but also to incorporate the cultural-evolutionary pronouncements of Leslie White (e.g., 1945, 1959a,b), Julian Steward (e.g., 1955), V. Gordon Childe (1951a), and their followers [e.g., Sahlins and Service 1960; Fried 1967; Service 1975; various papers in Rambo and Gillogly 1991 (cf. Rambo 1991)] into routine archaeological studies. Soon, archaeologists were taking Binford's exhortation seriously and were doing everything they could so as not to lose sight of the "Indian behind the artifact," as Robert Braidwood (1959:79), echoing Walter Taylor (1948), had so appropriately put it. Despite caveats raised by a few ethnographers—one of whom (Harris 1968:360) even encouraged archaeologists to "shrive yourselves of the notion that the units which you seek to reconstruct must match the units in social organization which contemporary ethnographers have attempted to tell you exist"—archaeologists began devising methods to determine whether the group that was responsible for "creating" a particular archaeological site was matrilineal or patrilineal (e.g., Deetz 1968; Longacre 1968; Hill 1970; Allen and Richardson 1971). These exercises were at first entertaining, but they began to lose some of the charm when holes began to appear in the anthropological armor in which archaeologists had clothed themselves. For one thing, such exercises were too particularistic. Although they might contribute tidbits of information that the ethnologists could use, the results were unsatisfying to the archaeologist, who wanted big answers to big questions. What about all of the regularities that ethnologists such as White said were there? How could *they* be found?

The answer was provided by Binford, who urged archaeologists to study the philosophy of science, which, he claimed (Binford 1972:17), he had been told to do by White. When the philosophy of science was then added to the equation—literally, when archaeologists were told that not only could one investigate culture process but also could do it *scientifically*—the stage was set for a mass exodus from the stifling constraints of such mundane pursuits as culture history. Now archaeology could get on with the exciting voyage of science, perhaps even discovering a few laws (empirical, covering, or otherwise) along the way. The self-described new archaeologists began paying homage to Carl Hempel and Ernest Nagel (see below)—in large part because Binford told them that was the correct thing to do—though few if any of them really understood such concepts as *hypothetico-deductive framework, deductive-nomological approach,* and *bridging arguments.* These simply were words that someone heard a philosopher or an archaeologist-turned-philosopher utter, and he or she was impressed because the words sounded scientific. News of the new and exciting terminology spread like wildfire, and soon an entire generation of archaeologists was (supposedly) doing science.

One highly influential book written during this halcyon period was *Explanation in Archaeology: An Explicitly Scientific Approach* (Watson *et al.* 1971), which proved to be so popular that is was revised and published over a decade later as *Archaeological Explanation: The Scientific Method in Archaeology* (Watson et al. 1984). Patty Jo Watson and her coauthors, Charles Redman and Steven LeBlanc, argued that archaeology could and should be a science and that one would know he or she had reached that goal if explanation was the end product of archaeological inquiry. To get there would require rigorous adherence to the scientific method: "Archaeologists should begin with clearly stated problems and then formulate testable hypothetical solutions. The degree of confirmation of conclusions should be exhibited by describing fully the field and laboratory data and the reasoning used to support these conclusions. This is what we mean by an explicitly scientific archaeological method" (Watson et al. 1984:129).

It is difficult to argue with the statement that archaeologists should state problems clearly and should describe data as completely as possible. And I find it difficult to argue against testing hypotheses, though technically what one actually is doing is examining the testable *implications* of a hypothesis. Watson et al. certainly were clear on their definition of science:

> ...science is based on the working assumption or belief by scientists that past and present regularities *are* pertinent to future events and that under similar circumstances similar phenomena will behave in the future as they have in the past and do in the present. This practical assumption of the regularity or conformity of nature is the necessary foundation for all scientific work. Scientific descriptions, explanations, and predictions all utilize lawlike generalizations hypothesized on the presumption that natural phenomena are orderly. (Watson et al. 1984:5–6)

> The ultimate goal of any science is construction of an axiomatized theory such that observed regularities can be derived from a few basic laws as premised. Such theories are used to explain past events and to predict future ones. Good theories lead to prediction of previously unsuspected regularities. Logical and mathematical axiomatic systems are essential as models of scientific theories, but no empirical science has yet been completely axiomatized. As Hempel indicates, it may ultimately turn out for any science, or for all sciences, that the goal is actually unattainable (Watson et al. 1984:14).

The Hempelian notion of science and how it operates formed the basis of the reintroduction into philosophy of 19th-century empiricism, though the term usually applied to Hempel's view is *logical positivism*. One of his books, *Aspects of Scientific Explanation, and Other Essays in the Philosophy of Science* (Hempel 1965), became as widely cited in the archaeological literature of the 1970s as it did in the philosophical literature of the 1960s, and it was his notion of science that Watson et al. (1971, 1984) assumed as the basis of their argument that archaeology could and should become scientific. For them, science was "an axiomatized theory such that observed regularities can be derived from a few basic laws as premises." Watson et al. were joined in their efforts to make the philosophy of science accessible to the archaeological community by other neopositivists such as Merrilee and Wesley Salmon (e.g., M. H. Salmon 1975, 1982; M. H. Salmon and W. C. Salmon 1979).

Central to all of their arguments was the Hempelian account of how one arrived at explanation—what Hempel termed the *deductive-nomological* approach. Despite his use of unusual terms, Hempel's basic tenet was simple: Whatever is to be explained (he used the term *explanandum* to refer to the "whatever") is derived logically from one or more universal statements, or laws, keeping in mind that certain boundary conditions might apply—hence, Watson and colleagues' (1984:5) above-cited definition of science as the "belief by scientists that past and present regularities *are* pertinent to future events and that under similar circumstances similar phenomena will behave in the future as they have in the past and do in the present."

There are still a few philosophers around who view science in Hempelian terms, but by the middle of the 1970s it was becoming clear that the deductive-nomological approach was dying a natural death. There were attempts to keep it alive, for example by linking it to the *bridging-law* concept of philosopher Ernest Nagel (e.g., 1961), but these also died out—except among archaeologists, who began making bridges between the archaeological present and the archaeological past through such things as ethnographic analogy and ethnoarchaeology (see Fritz 1972). In other words, archaeologists were using the present as an analogue of the past. In fact, they *had* to resort to analogy; how else were they going to find the laws that Hempel said were there—the very laws that, once discovered, led to the formulation of "axiomatized theory" and thus ultimately to explanation?

What a blessing it was that archaeologists now had access to the past through the present. They could find patterning in their archaeological data sets and interpret the patterning in terms of modern analogues. Or, conversely, they could use present behavior as a guide to what to look for in the prehistoric archaeological record. If one found enough correlations between the past and the present, then surely laws could be constructed to account for the similarity in pattern. Any slight deviations could be explained away in terms of slightly different "boundary conditions," to use Hempel's term, that had impinged on the creators of the past and present signatures. The end result of this exercise was scientific explanation—defined as interpretation by way of law formulation. This is the reason why Watson (1986:452) equates archaeological interpretation with "describing and explaining the real past."

There are, however, several archaeologists, myself included, who do not agree with this conflation of interpretation and explanation nor with the belief that the Hempelian view of science can be applied to the study of organisms, including humans. The type of science Watson has in mind—a predictive, law-driven science—will not work in archaeology. Hempelian science is not particularly useful for studying humans—or any other organism—because of the assumptions it makes relative to laws. No one has ever denied that chemical–physical laws do not apply to organisms, but at the level that concerns most archaeologists—behavior (why we do what we do) and the products of those behaviors (artifacts)—they do not appear to play a deterministic role. And determinism, i.e., the intrinsic properties that something has that makes that something predictable, is the basis of Hempelian explanation (again, within reason—remember Hempel's "boundary conditions").

Deterministic laws work well for physical things such as elements and molecules and their chemical interactions, but they do not work well for organisms. A carbon atom, for example, is always a carbon atom, regardless of time or place. And there are deterministic laws that govern how carbon atoms interact with other atoms. For example, if four hydrogen atoms happen to pass near a carbon atom, it is a safe bet that the carbon atom will grab them and form a molecule of methane. We can make that bet today, tomorrow, or 10 years from now and we will win it—just as Hempelian science says we will. The safety of the bet resides in our knowing what the laws are that govern the behaviors of atoms *and* in our understanding the various chemical–physical mechanisms that carry out the dictates of the laws. Those kinds of laws apply the invariant properties of *inanimate* objects, but they do not work on such things as the behavior of organisms (O'Brien and Holland 1990, 1995b). They are, however, precisely the kinds of laws archaeologists wanted to apply to humans. Fritz and Plog (1970:405), for example, were explicit about the definition of law: "A statement of relation between two or more variables which is true *for all times and places*" (italics added).

One could, I suppose, dance around the issue and claim a distinction between "universal facts" and "laws"—Binford (1972:18) claimed that Leslie White once noted that "Julian Steward doesn't know the difference between a universal fact and a law" (I'm not sure I do either)—but this obscures the real issue, namely, are there invariant laws that govern human behavior? If there are, then the Hempelian notion of science is quite adequate. If there are not, then where do we look for explanation? We might start by looking at scientific evolutionary theory, which has little or nothing to do with invariant laws.

EVOLUTIONARY THEORY

The new archaeologists forgot (or never knew) that invariant laws are not the only kind of laws around. What about the law of contingency, which says that whatever happens at point D is conditioned in part by what happened at points A, B, and C? Point D is not *determined* by what happened at the other points but rather is *contingent* on what happened at those points. Whatever is manifest at point D is *stochastic* as opposed to *random*—meaning that the expression of D is derived from a limited, rather than an *infinite*, number of possibilities. The theory of Darwinian evolution is built around contingency as a historical process. Organisms evolve, and in the process usually change shape, but there is not an unlimited number of shapes into which they can change from generation to generation. There are certain forms that hang on for hundreds, thousands, or millions of generations, sometimes with little variation evident from one generation to the next. In other words, forms are *channeled* in certain directions because of their history (Mayr 1988:108). We do not, for example, expect a newborn baby to have three legs. It might have six toes—that's not too uncommon—but an extra leg is almost impossible to imagine. Few of those knowledgeable in the natural sciences would argue the Darwinian evolution is not a theory or that in its modified form (modified in the sense that we now understand genetic transmission, embryonic development, and the like) it is not capable of providing *explanations* for how and why we are the way we are. It is not a perfect theory—yet—but it is a good one because it works.

But, as Holland and I have pointed out (O'Brien and Holland 1995a), a review of the history of scientific paradigms makes it clear that for any paradigm to take hold requires a considerable amount of time and reiteration. To say that evolutionary archaeology is a good paradigm because it works really says nothing about *how* and, more importantly, *why* it works. Those two issues—the how and why it works—can only be addressed through a careful reading of the biological, not the archaeological, literature. Application of the theory to archaeological phenomena is entirely appropriate, and it is equally clear that archaeology has something to contribute to evolutionary theory. But the theory itself is a biological one, not an archaeological one.

I personally do not find the fact that Darwinian evolution has been applied to the archaeological record particularly novel. I suspect that it was only a matter of time before archaeologists began to see the archaeological record for what it is—a record of the histories of past human phenotypes. It is rather surprising, however, that anthropologists, as opposed to archaeologists, were not the first to seize on the idea that Darwinian evolution is entirely appropriate to the study of all humans, their behaviors, and their behavioral by-products. The notion that selection, the centerpiece of Darwinian evolution, operates on humans might have been profound back in 1859, but after Darwin published *On the Origin of Species*, it theoretically should have been a relatively uncomplicated matter to extrapolate "descent with modification" to humans and, by extension, to features that affect their fitness. However, this extrapolation was slow to be made. Not even Darwin wanted to admit that humans were necessarily a product of natural selection and other evolutionary processes (many of which were unknown or misunderstood), a view that still pervades anthropology and inhibits the acceptance of an internally consistent approach to the study of humans and the materials they manufacture, use, and discard. Evolutionary archaeology, however, has made the claim that humans, like any other organisms, are directly affected by selection and that some aspects of the material record reflect the effects of selection (Dunnell 1982; Leonard and Jones 1987; O'Brien and Holland 1990, 1992, 1995a; O'Brien et al. 1994).

Evolution, of course, has been around in Americanist archaeology for a long time. In the late 19th century it was evolution as espoused by Lewis Henry Morgan and others to which most Bureau of American Ethnology prehistorians and ethnologists subscribed, at least in part. Later, as we have seen, it was evolution according to White, Steward, and friends that caught the attention of archaeologists. Interspersed were the purely functional arguments of anthropologists such as Marvin Harris (e.g., 1979) and the aforementioned systems-theory formulations of archaeologists such as Flannery. Evolution often was invoked in such formulations, though it bore little or no resemblance to anything familiar to biologists. For the most part, anthropological brands of evolution were and still are little more than unidirectional, progressive formulations grounded in the notion of some kind of cultural transformation (Dunnell 1980; Rambo 1991). Change is viewed simplistically as an outcome of need. For example, if a group is facing food shortage, it simply forms alliances with other groups to develop a different means of obtaining food. In most anthropological schemes, groups (and, by extension, individuals within the groups) always come out as winners.

Selection, other than some vague notion of *cultural* selection, plays no role in most evolutionary scenarios concocted by anthropologists and archaeologists, since evolution becomes little more than a set of invented solutions to problems posed by the environment (Lewontin 1983). In other words, humans go out and get whatever it is they need to adapt to their social and physical environment. In a real sense, anthropologists emphasize humans as intent-driven, *maximizing* creatures, a concept that has been amplified in anthropology through the addition of sociobiology as an area of interest. Why should we believe that humans act any differently than other organisms when it comes to behaviors? Certainly there is nothing in evolutionary theory that states that organisms must always act in accordance with some maximizing strategy. As Dawkins (1990:188–189) notes, "Individuals do not consciously strive to maximize anything; they behave *as if* maximizing something....individuals may strive for something, but it will be a morsel of food, an attractive female, or a desirable territory." As Darwin himself figured out, no such thing as a perfectly adapted organism has existed or will ever exist. All he ever had in mind when he used the phrase "survival of the fittest" was for "the *tendency* of organisms that are better engineered to be reproductively successful" (Burian 1983:299; italics added). In other words, "If a is better adapted than b in environment E, then (probably) a will have greater reproductive success than b in E" (Brandon 1990:11). The kinds of "explanations" that usually result from mechanistic application to humans of concepts such as optimal foraging strategy are not science, they're just-so stories (Dunnell and Wenke 1979).

A single example should serve to demonstrate not only the way intentionality is interwoven in adaptationist stories but also the pervasiveness of the adaptationist perspective in archaeology. Some of the greatest just-so stories in archaeology have centered around the origin(s) of agriculture, as if domestication and attendant processes are the result of, to borrow a phrase from Childe (1951b), man's attempt to make himself. In other words, agriculture is viewed as a solution to an environmental problem, be it population pressure (Cohen 1975, 1977) or a host of other problems. In two works, David Rindos (1980, 1984) provided a clear exposition of how Darwinian evolutionary theory can, in essence, explain the origin and spread of domesticatory systems. Importantly, his explanation says nothing about human intent and invention, a fact he points out explicitly: "Parsimony would suggest that if agricultural origins may be explained without the use of intent or invention, then these concepts may, for the purposes of this model, be set aside" (Rindos 1980:751). As might have been anticipated, not all anthropologists and archaeologists were kind in their assessments of Rindos's evolutionary explanation (e.g., Ceci 1980; Shaffer 1980; Yarnell 1985; Flannery 1986), pointing out repeatedly that any "model" of the origins of agriculture *must* take into account human intention and problem-solving abilities.

It is not going to be a simple matter to eradicate storytelling from archaeology, and, in fact, without a real understanding of Darwinian evolution and its attendant processes such as selection and drift, we run the risk of substituting evolutionary-based "adaptationist" stories for the orthogenetic ones stemming from cultural evolution. In other words, we cannot assume that by wrapping ourselves in Darwin's mantle that our stories are any better than those from someone wrapped in the mantle of White or Steward. Silly adaptationist stories are as much a problem in archaeology (O'Brien and Holland 1990, 1992, 1995a) as they are in biology (Gould and Lewontin 1979) and, importantly, pose a serious threat to the profession taking evolutionary archaeology seriously. These can be minimized, especially through reliance on engineering-design analysis (Mayr 1983; O'Brien and Holland 1990, 1992; O'Brien et al. 1994).

DISCUSSION

It might appear that the amount of time that has passed since the first seeds of an evolutionary archaeology were planted—I use 1978 as a benchmark, for it was in that year that Dunnell (1978a,b,c) sketched out in three papers some of the essential points of such an approach—is an inordinately long time for an approach to have been around with few if any takers. Despite the theoretical and methodological advances that have been made in the following decade and a half (e.g., Dunnell 1982, 1985, 1988, 1989b, 1992, 1995; Leonard and Jones 1987; O'Brien and Hollard 1990, 1992, 1995a,b; Neff 1992; Teltser 1995), critics might argue that even now the number of actual case studies that employ evolutionary theory is small. Proponents of the approach might take heart in knowing that as late as the middle 1930s it was by no means clear to biologists exactly how Darwinian evolution worked. On one side were the naturalists and their ideas on geographic isolation as a major cause of speciation. On the other side were the experimentalists and their ideas on mutationism. The gulf between gradual evolution by means of natural selection and rapid evolution by means of mutation seemed unbridgeable, but by the middle 1930s the situation changed dramatically. As Mayr (1982:566–567) points out, two things had to happen before a bridge could be constructed: (1) geneticists had to take an interest in both diversity and the populational aspects of evolution and (2) naturalists had to understand that the experimentalists (geneticists) no longer were opposed to natural selection and gradualism. The latter group also had to abandon its emphasis on the transmission of acquired characteristics. Within about a decade, biologists reached what Huxley (1942) termed the *evolutionary synthesis*.

I imagine the same thing will happen in archaeology. As archaeologists become more familiar with evolutionary theory and begin to move outside their narrow specialties, applications will grow exponentially. If we can escape the temptation to construct patently absurd adaptationist scenarios that ostensibly "explain" variation in the archaeological record, evolutionary archaeology will become widely accepted as a legitimate approach. I take sharp exception with those who note with derision that the number of case studies in evolutionary archaeology is still so small after all these years, as if this is evidence that somehow the approach is flawed. There is nothing flawed with the approach; what is flawed is our thinking. It is still difficult for many people to believe that selection works on humans, as if the fact that we have "culture" somehow makes us immune to selection and drift. This is patently nonsense. And neither do we have to invoke a special kind of selection—"cultural selection"—to address the issue of human evolution. Selection does not need to be gussied up in new clothes for application to humans. Neither do we need to be worried at this stage about the type of vehicle by which variation is transmitted or how the variation arose. Selection, in fact, is blind to the source of variation (O'Brien and Hollard 1990), and all that matters is that the variation is present and that it can be transmitted. Humans might have a few more cards with which to play the game than other animals do, but the rules are the same. I suggest that instead of searching for a separate set of rules, which does not exist, archaeologists should examine who has won and lost the games played over the last 10,000 years or so and attempt to figure out *how* and *why* the winners won and the losers lost. Evolutionary theory offers a means of doing this without recourse to inventing little stories.

Above all, archaeologists who want to make the discipline scientific need to keep firmly in mind that the type of science to which they aspire is historical as opposed to physical in nature. In physical science, prediction is symmetrical to causation (Mayr 1982:71); in historical science there is no prediction. Thus, attention spent on law formulation in archaeology is pointless, since there cannot *be* any laws except that of contingency. Rather than search for "explanation" in terms of "universal facts" and "laws," archaeologists should realize that explanation is derived from the theory itself. Mayr's (1982:76) admonition to biologists is equally appropriate for archaeologist: "what is needed is an uncommitted philosophy of biology which stays equally far away from vitalism and other unscientific ideologies and from a physicalist reductionism that is unable to do justice to specifically biological phenomena and systems."

REFERENCES

Allen, W.L., and Richardson, J.B., III, 1971, The Reconstruction of Kinship From Archaeological Data: The Concepts, the Methods, and the Feasibility, *American Antiquity* 36:41–53.
Bennett, J.W., 1943, Recent Developments in the Functional Interpretation of Archaeological Data, *American Antiquity* 8:208–219.
Binford, L.R., 1962, Archaeology as Anthropology, *American Antiquity* 28:217–225.
Binford, L.R., 1968, Archeological Perspectives, in: *New Perspectives in Archeology* (S.R. Binford and L.R. Binford, eds.), Aldine Press, Chicago, pp. 5–32.
Binford, L.R., 1972, *An Archaeological Perspective*, Seminar Press, New York.
Binford, S.R., and Binford, L.R. (eds.), 1968, *New Perspectives in Archeology*, Aldine Press, Chicago.
Braidwood, R.J., 1959, Archaeology and the Evolutionary Theory, in: *Evolution and Anthropology: A Centennial Appraisal* (B.J. Meggers, ed.), The Anthropological Society of Washington, Washington, DC, pp. 76–89.
Brandon, R.N., 1990, *Adaptation and Environment*, Princeton University Press, Princeton.
Braun, D.P., 1990, Selection and Evolution in Nonhierarchical Organization, in: *The Evolution of Political Systems: Sociopolitics in Small-Scale Sedentary Societies* (S. Upham, ed.), Cambridge University Press, Cambridge, pp. 62–86.
Burian, R., 1983, Adaptation, in: *Dimensions of Darwinism* (M. Grene, ed.), Cambridge University Press, Cambridge, pp. 287–314.
Caldwell, J.C., 1958, The New American Archaeology, *Science* 129:303–307.
Ceci, L., 1980, Comment on "Symbiosis, Instability, and the Origins and Spread of Agriculture," by D. Rindos, *Current Anthropology* 21:766.
Childe, V.G., 1951a, *Social Evolution*, Schuman, New York.
Childe, V.G., 1951b, *Man Makes Himself*, Mentor, New York.
Cohen, M.N., 1975, Population Pressure and the Origin of Agriculture, in: *Population, Ecology, and Social Evolution* (S. Polgar, ed.), Mouton, The Hague, pp. 79–121.
Cohen, M.N., 1977, *The Food Crisis in Prehistory: Overpopulation and the Origins of Agriculture*, Yale University Press, New Haven.

Dawkins, R., 1990, *The Extended Phenotype: The Long Reach of the Gene*, 2 Ed., Oxford University Press, Oxford.
Deetz, J., 1968, The Inference of Residence and Descent Rules from Archeological Data, in: *New Perspectives in Archeology* (S.R. Binford and L.R. Binford, eds.), Aldine Press, Chicago, pp. 41–48.
Dobzhansky, T., 1951, Mendelian Populations and Their Evolution, in: *Genetics in the 20th Century* (L.C. Dunn, ed.), Macmillan Co., New York, pp. 573–589.
Dunnell, R.C., 1978a, Style and Function: A Fundamental Dichotomy, *American Antiquity* 43:192–202.
Dunnell, R.C., 1978b, Archaeological Potential of Anthropological and Scientific Models of Function, in: *Archaeological Essays in Honor of Irving B. Rouse* (R.C. Dunnell and E.S. Hall, Jr., eds.) Mouton, The Hague, pp. 41–73.
Dunnell, R.C., 1978c, Natural Selection, Scale, and Cultural Evolution: Some Preliminary Considerations, Paper presented at the 77th Annual Meeting of the American Anthropological Association, Los Angeles.
Dunnell, R.C., 1980, Evolutionary Theory and Archaeology, in: *Advances in Archaeological Method and Theory*, Vol. 3 (M.B. Schiffer, ed.), Academic Press, New York, pp. 35–99.
Dunnell, R.C., 1982, Science, Social Science, and Common Sense: The Agonizing Dilemma of Modern Archaeology, *Journal of Anthropological Research* 38:1–25.
Dunnell, R.C., 1985, Methodological Issues in Contemporary Americanist Archaeology, in: *Proceedings of the 1984 Biennial Meeting of the Philosophy of Science Association*, Vol. 2 (P.D. Asquith and P. Kitcher, eds.), Philosophy of Science Association, East Lansing, MI, pp. 717–744.
Dunnell, R.C., 1986, Methodological Issues in Americanist Artifact Classification, in: *Advances in Archaeological Method and Theory*, Vol. 9 (M.B. Schiffer, ed.), Academic Press, New York, pp. 149–207.
Dunnell, R.C., 1988, The Concept of Progress in Cultural Evolution, in: *Evolutionary Progress* (M.H. Nitecki, ed.), University of Chicago Press, Chicago, pp. 169–194.
Dunnell, R.C., 1989a, Philosophy of Science and Archaeology, in: *Critical Traditions in Contemporary Archaeology* (V. Pinsky and A. Wylie, eds.), Cambridge University Press, Cambridge, pp. 5–9.
Dunnell, R.C., 1989b, Aspects of the Application of Evolutionary Theory in Archaeology, in: *Archaeological Thought in America* (C.C. Lamberg-Karlovsky, ed.), Cambridge University Press, Cambridge, pp. 35–49.
Dunnell, R.C., 1992, Archaeology and Evolutionary Science, in: *Quandaries and Quests: Visions of Archaeology's Future* (L. Wandsnider, ed.), Occasional Paper No. 20, Center for Archaeological Investigations, Southern Illinois University, Carbondale, pp. 209–224.
Dunnell, R.C., 1995, What Is It That Actually Evolves?, in: *Evolutionary Archaeology: Methodological Issues* (P. A. Teltser, ed.), University of Arizona Press, Tucson, pp. 33–50.
Dunnell, R.C., and Leonard, R.D. (eds.), n.d., *Archaeological Applications of Evolutionary Theory*.
Dunnell, R.C., and Wenke, R.J., 1979, An Evolutionary Model of the Development of Complex Society, Paper presented at the Annual Meeting of the American Association for the Advancement of Science, San Francisco.
Feathers, J.K., 1989, Ceramic Analysis, Variation and Database Construction: A Selectionist Perspective, in: *Analysis and Publication of Ceramics: The Computer Data-Base in Archaeology* (J.A. Blakely and W.J. Bennett, Jr., eds.), British Archaeological Reports International Series No. 551, pp. 71–80.
Flannery, K.V., 1968a, Archaeological Systems Theory and Early Mesoamerica, in: *Anthropological Archaeology in the Americas* (B.J. Meggers, ed.), Anthropological Society of Washington, Washington, DC, pp. 67–87.

Flannery, K.V., 1968b, The Olmec and the Valley of Oaxaca: A Model of Interregional Interaction in Formative Times, in: *Dunbarton Oaks Conference on the Olmec* (E.P. Benson, ed.), Dunbarton Oaks, Washington, DC, pp. 79–110.
Flannery, K.V., 1972, The Cultural Evolution of Civilizations, *Annual Review of Ecology and Systematics* 3:399–426.
Flannery, K.V., 1986, A Visit to the Master, in: *Guila Naquitz: Archaic Foraging and Early Agriculture in Oaxaca, Mexico* (K.V. Flannery, ed.), Academic Press, New York, pp. 511–519.
Flannery, K. V., and Coe, M.D., 1968, Social and Economic Systems in Formative Mesoamerica, in: *New Perspectives in Archaeology* (S.R. Binford and L.R. Binford, eds.), Aldine Press, Chicago, pp. 267–283.
Fried, M.H., 1967, *The Evolution of Political Society*, Random House, New York.
Fritz, J.M., 1972, Archaeological Systems from Indirect Observation of the Past, in: *Contemporary Archaeology: A Guide to Theory and Contributions* (M.P. Leone, ed.), Southern Illinois Press, Carbondale, pp. 135–157.
Fritz, J.M., and Plog, F.T., 1970, The Nature of Archaeological Explanation, *American Antiquity* 35:405–412.
Ghiselin, M.T., 1966, On Psychologism in the Logic of Taxonomic Controversies, *Systematic Zoology* 15:207–215.
Ghiselin, M.T., 1974, A Radical Solution to the Species Problem, *Systematic Zoology* 25:536–544.
Ghiselin, M.T., 1981, Categories, Life and Thinking, *Behavioral and Brain Sciences* 4:269–283.
Gould, S.J., and Lewontin, R., 1979, The Spandrels of San Marco and the Panglossian Paradigm: A Critique of the Adaptationalist Program, *Proceedings of the Royal Society of London* B205:581–598.
Haag, W.G., 1959, The Status of Evolutionary Theory in American Archaeology, in: *Evolution and Anthropology: A Centennial Appraisal* (B.J. Meggers, ed.), The Anthropological Society of Washington, Washington, DC, pp. 90–105.
Harris, M., 1968, Comments by Marvin Harris, in: *New Perspectives in Archeology* (S.R. Binford and L.R. Binford, eds.), Aldine Press, Chicago, pp. 359–361.
Harris, M., 1979, *Cultural Materialism: The Struggle for a Science of Culture*, Random House, New York.
Hempel, C.G., 1965, *Aspects of Scientific Explanation, and Other Essays in the Philosophy of Science*, Free Press, New York.
Hill, J.N., 1970, *Broken K Pueblo: Prehistoric Social Organization in the American Southwest*, Anthropological Papers of the University of Arizona No. 18.
Huxley, J.S., 1942, *Evolution, the Modern Synthesis*, Allen and Unwin, London.
Leonard, R.D., and Jones, G.T., 1987, Elements of an Inclusive Evolutionary Model for Archaeology, *Journal of Anthropological Archaeology* 6:199–219.
Lewontin, R.C., 1983, The Organism as the Subject and Object of Evolution, *Scientia* 118:65–82.
Longacre, W.A., 1968, Some Aspects of Prehistoric Society in East-central Arizona, in: *New Perspectives in Archeology* (S.R. Binford and L.R. Binford, eds.), Aldine Press, Chicago, pp. 89–102.
Mayr, E., 1963, *Animal Species and Evolution*, Harvard University Press, Cambridge, MA.
Mayr, E., 1973, *Populations, Species, and Evolution*, Harvard University Press, Cambridge, MA.
Mayr, E., 1976, Typological versus Population Thinking, in: *Evolution and the Diversity of Life*, Harvard University Press, Cambridge, MA, pp. 26–29.
Mayr, E., 1977, Darwin and Natural Selection, *American Scientist* 65:321–337.
Mayr, E., 1982, *The Growth of Biological Thought*, Harvard University Press, Cambridge, MA.
Mayr, E., 1983, How to Carry Out the Adaptationist Program?, *The American Naturalist* 121:324–334.
Mayr, E., 1987, The Ontological Status of Species, *Biology and Philosophy* 2:145–166.

Mayr, E., 1988, *Toward a New Philosophy of Biology: Observations of an Evolutionist*, Harvard University Press, Cambridge, MA.
Nagel, E., 1961, *The Structure of Science*, Harcourt, Brace & World, New York.
Neff, H., 1992, Ceramics and Evolution, in: *Advances in Archaeological Method and Theory*, Vol. 2 (M.B. Schiffer, ed.), University of Arizona Press, Tucson, pp. 141–194.
O'Brien, M.J., 1996, *Evolutionary Archaeology: Theory and Applications*. University of Utah Press, Salt Lake City.
O'Brien, M.J., and Holland, T.D., 1990, Variation, Selection, and the Archaeological Record, in: *Advances in Archaeological Method and Theory*, Vol. 2 (M.B. Schiffer, ed.), University of Arizona Press, Tucson, pp. 31–79.
O'Brien, M.J., and Holland, T.D., 1992, The Role of Adaptation in Archaeological Explanation, *American Antiquity* 57:36–59.
O'Brien, M.J., and Holland, T.D., 1995a, The Nature and Premise of a Selection-based Archaeology, in: *Evolutionary Archaeology: Methodological Issues* (P.A. Teltser, ed.), University of Arizona Press, Tucson, pp. 175–200.
O'Brien, M.J., and Holland, T.D., 1995b, Behavioral Archaeology and the Extended Phenotype, in: *Expanding Archaeology* (J.M. Skibo, W.H. Walker, and A.E. Nielson, eds.), University of Utah Press, Salt Lake City, pp. 143–161.
O'Brien, M.J., Holland, T.D., Hoard, R.J., and Fox, G.L., 1994, Evolutionary Implications of Design and Performance Characteristics of Prehistoric Pottery, *Journal of Archaeological Method and Theory* 1:259–304.
Rambo, A.T., 1991, The Study of Cultural Evolution, in: *Profiles in Cultural Evolution* (A.T. Rambo and K. Gillogly, eds.), Anthropological Papers No. 85, Museum of Anthropology, University of Michigan, Ann Arbor, pp. 23–109.
Rambo, A.T., and Gillogly, K. (eds.), 1991, Profiles in Cultural Evolution, Anthropological Papers No. 85, Museum of Anthropology, University of Michigan, Ann Arbor.
Rindos, D., 1980, Symbiosis, Instability, and the Origins and Spread of Agriculture: A New Model, *Current Anthropology* 21:751–772.
Rindos, D., 1984, *The Origins of Agriculture: An Evolutionary Perspective*, Academic Press, New York.
Rindos, D., 1989, Undirected Variation and the Darwinian Explanation of Cultural Change, in: *Advances in Archaeological Method and Theory*, Vol. 1 (M.B. Schiffer, ed.), University of Arizona Press, Tucson, pp. 1–45.
Sahlins, M.D., and Service, E.R. (eds.), 1960, *Evolution and Culture*, University of Michigan Press, Ann Arbor.
Salmon, M.H., 1975, Confirmation and Explanation in Archeology, *American Antiquity* 40:459–464.
Salmon, M.H., 1982, *Philosophy and Archaeology*, Academic Press, New York.
Salmon, M.H., and Salmon, W.C., 1979, Alternative Models of Scientific Explanation, *American Anthropologist* 81:61–74.
Service, E.R., 1975, *Origins of the State and Civilization*, Norton, New York.
Shaffer, J.G., 1980, Comment on "Symbiosis, Instability, and the Origins and Spread of Agriculture," by D. Rindos, *Current Anthropology* 21:768.
Sober, E., 1980, Evolution, Population Thinking, and Essentialism, *Philosophy of Science* 47:350–383.
Sober, E., 1984, *The Nature of Selection: Evolutionary Theory in Philosophical Focus*, MIT Press, Cambridge, MA.
Steward, J.H., 1942, The Direct Historical Approach to Archaeology, *American Antiquity* 7:337–343.

Steward, J.H., 1955, *Theory of Culture Change: The Methodology of Multilinear Evolution*, University of Illinois Press, Urbana.
Steward, J.H., and Setzler, F.M., 1938, Function and Configuration in Archaeology, *American Antiquity* 4:4–10.
Taylor, W.W., 1948, *A Study of Archeology*, American Anthropological Association, Memoir No. 69.
Teltser, P.A. (ed.), 1995, *Evolutionary Archaeology: Methodological Issues*, University of Arizona Press, Tucson.
Watson, P.J., 1986, Archaeological Interpretation, 1985, in: *American Archaeology Past and Future: A Celebration of the Society for American Archaeology, 1935–1985* (D.J. Meltzer, D.D. Fowler, and J.A. Sabloff, eds.), Smithsonian Institution Press, Washington, DC, pp. 439–457.
Watson, P.J., LeBlanc, S.A., and Redman, C., 1971, *Explanation in Archeology: An Explicitly Scientific Approach*, Columbia University Press, New York.
Watson, P.J., LeBlanc, S.A., and Redman, C., 1984, *Archeological Explanation: The Scientific Method in Archaeology*, Columbia University Press, New York.
White, L.A., 1945, History, Evolution, and Functionalism: Three Types of Interpretation of Culture, *Southwestern Journal of Anthropology* 1:221–248.
White, L.A., 1959a, The Concept of Evolution in Anthropology, in: *Evolution and Anthropology: A Centennial Appraisal* (B.J. Meggers, ed.), The Anthropological Society of Washington, Washington, DC, pp. 106–124.
White, L.A., 1959b, *The Evolution of Culture*, McGraw–Hill, New York.
Willey, G.R., 1962, The Early Great Styles and the Rise of the Pre-Columbian Civilizations, *American Anthropologist* 64:1–14.
Willey, G.R., and Phillips, P., 1958, *Method and Theory in American Archaeology*, University of Chicago Press, Chicago.
Yarnell, R., 1985, Review of *The Origins of Agriculture: An Evolutionary Perspective*, by D. Rindos, *American Antiquity* 50:698–699.

Chapter 3

Explaining the Change from Biface to Flake Technology
A Selectionist Application

ALYSIA L. ABBOTT, ROBERT D. LEONARD, AND
GEORGE T. JONES

INTRODUCTION

Why do Coke bottles have ridges? Why did the "American system" of manufacture, with its production of identical, interchangeable parts, fail in the homeland of its invention, France? Why do all of the great American cars of the 1950s have tail fins? Why were Mississippi riverboats created for a working life of 3 or 4 years? Why do home blenders have between 1 and 18 speeds? Why the Qwerty rather than the more "logical" Dvorak keyboard? Why were there over 800 tractor manufacturers in the early part of this century, and only a handful today? Why is the geared eggbeater the standard in American kitchens, while Europeans continue to beat eggs with a whisk? Why did Thomas Jefferson's perfectly designed plow, the "Mouldboard of Least Resistance," win an award from the American Philosophical Society, but not the acceptance of the American farmer?

Depending on who is doing the ranking, the questions posed above (many inspired by the fascinating book *Made in USA* by Phil Patton) might be

---

ALYSIA L. ABBOTT AND ROBERT D. LEONARD • Department of Anthropology, University of New Mexico, Albuquerque, New Mexico 87131.   GEORGE T. JONES • Department of Anthropology, Hamilton College, Clinton, New York 13323.

considered as ranging from quite significant to most trivial. Many are certainly mundane (in the most generous sense of the word), ostensibly of interest only to historians of the plow or chroniclers of 19th- and 20th-century kitchen gadgetry. While considerations of blending and egg-beating technology may be mundane, it is intellectually self-defeating for any archaeologist to conclude that this question or any other posed above is irrelevant or trivial, since each constitutes an example of the same kinds of questions archaeologists pose concerning prehistoric technologies: Why do some technologies replace others? Why are the perceived "best" designs frequently not the most used? Why do so many technologies exist to serve the same function?

The common thread running through these technological questions is that they are not merely historical, they are also evolutionary. Humans have evolutionary histories with which our behaviors, and the technological products of those behaviors, are inexorably linked. These evolutionary histories *must* be explained in evolutionary terms. Importantly, we must recognize that the form, the distribution, the success of all technologies—from eggbeaters, to Stealth fighters, to ceramics, to bifaces—are shaped by the same basic evolutionary processes.

The purpose of this paper is to consider these evolutionary processes in providing an explanation for the shift from biface- to flake-based technologies that occurred in many areas of North America in prehistory. The evolutionary framework we employ is Darwinian, and has been termed *selectionist theory*. As processual archaeologists have considered this shift previously (e.g., Torrence 1983; Parry and Kelly 1987; Jeske 1992; Sassaman 1992), our discussion here also constitutes a contrast between processual and selectionist thought.

THE STRUCTURE OF SELECTIONIST THOUGHT

While a detailed discussion of the selectionist framework is beyond the scope of this paper, fundamental concepts need to be introduced (see Dunnell 1980, 1989; Leonard and Jones 1987; Leonard 1989; Rindos 1989; Braun 1990; O'Brien and Holland 1990, 1992; Dunnell and Feathers 1991; Leonard and Abbott 1993, for examples of selectionist theory and applications). The major premises underlying the selectionist perspective are that human behavior, and as a consequence technology, are components of the human phenotype, and that phenotypic change may be explained by the operation of natural selection on behavioral variants, including technological behaviors. When considering technological variants, variation of interest is described, and the *replicative success* of variants documented through time. For the most part, changes in the shape of distributions are argued to be either stochastic (style) or directional, with directionality being either the consequence of the operation of natural selection on that variant (which makes it functional [*sensu* Dunnell 1978], and hence adaptive), or the invariable association of one variant that is not necessarily functional with one that is (sorting). When directionality is observed, the identification of the *selective agents* on the variant or the variant it is sorted with constitutes evolutionary explanation.

Natural selection is seen as the mechanism of change, and importantly, technology is not assumed, *a priori*, to be of adaptive significance as in most processual thought. As a result, adaptations are identified *only* after the operation of natural selection has been demonstrated. Selectionist theory differs from processual thought here, as most processualist studies regard adaptation as the mechanism for change, and assume technology to be reflexive of adaptation. Minimally, in processual thought, technology is presumed, *a priori*, to be an active component of an "adaptive system." Whereas processualism views technology as the consequences of people acting to improve their adaptation, selectionism views technology as comprised of variants that can be seen as competing alternatives that have different consequences for users, and hence their own replication. Gould and Vrba (1982), in a review of the concept of adaptation, argue that in order for traits to be considered adaptations, they must contribute to current fitness *and* be the product of natural selection. We therefore can neither assume that a particular technology was the product of selection (it may be neutral [stylistic], or merely sorted [invariably assorted with something that *is* selected]), nor that it has impacted the reproductive success of the bearers of that technology on average. Even if we can make a reasonable argument that a particular technology is an adaptation, or a component of an adaptation (e.g., a class of projectile points) we still do not necessarily know how that "adaptation" contributed to fitness, i.e., how the bearers of this technology were at a reproductive advantage over those who utilized alternative technologies.

Herein lies one of the most fundamental theoretical differences between selectionist and processual thought. *Selectionist theory demands that explanations consider these complex relationships among traits with respect to neutrality, sorting, selection, and adaptation.* Processual thought lacks these theoretical components. Our consideration of the shift from a biface- to flake-based technology illustrates this difference, highlighting the potential of selectionist theory.

THE TRANSITION FROM BIFACE TO FLAKE TECHNOLOGY: A PROCESSUAL EXPLANATION

The shift from biface- to flake-based technology has been documented by a number of archaeologists in the archaeological records of the Eastern woodlands (Montet-White 1968; Hofman and Morrow 1985), Mesoamerica (Flannery et al. 1981; Parry 1983), the North American Southwest (Parry and Christenson 1986), and the Great Plains (Parry and Kelly 1987). Empirical support for the presence of the shift is based on a documented increase in the relative frequency of generalized cores and utilized flakes with a corresponding decrease in the relative frequency of bifaces as well as the debitage associated with bifacial reduction (Parry and Kelly 1987). While there has been considerable work toward providing an explanation for this apparent shift from one technological organization to another, the most widely accepted explanation has been put forth by Parry and Kelly (1987)

The structure of their processual argument is as follows:

1. *A pattern is observed in the shift from biface to flake production in many areas of North America during prehistory.* Parry and Kelly note that there is empirical evidence to support a shift from bifacial reduction to generalized core reduction demonstrable in the archaeological record of several regions of North America.
2. *Ethnographic descriptions of contemporary peoples practicing unstandardized core reduction are presented.* Ethnographic descriptions are used here in order to establish behavioral correlations between unstandardized core reduction and the archaeological record.
3. *Commonalities are sought in the ethnographic descriptions.* This section incorporates generalities in the observed behavior of unstandardized core reduction in order to compile the "characteristics" of the technology. The resulting list of unstandardized core reduction characteristics is as follows:

    a. "the flaking techniques are not intended to control the form of the resulting flakes" (p. 287).
    b. "no explicit distinction is made between "tools" and "waste" (p. 287).
    c. "the tools are seldom modified" (p. 287).

4. *Conclusions are drawn as to the nature of the shift in the archaeological record.* The ethnographically derived characteristics of unstandardized core reduction are concluded to be indicative of an expedient technology. Application of these characteristics to the prehistoric assemblages then allows Parry and Kelly to conclude that the shift: "may be viewed as a change from a curated to a more expedient technology" (p. 288).
5. *Archaeological correlates with expedient core technology are sought.* Given the conclusion drawn in step #4 that the shift from a standardized to an unstandardized technology is representative of a shift from a curated to an expedient technology, Parry and Kelly seek to find potential causes for the shift in the form of correlations with environ-

mental factors, topography, and raw material procurement. No correlations were found with the availability of raw material, topography, climate, vegetation, or other set of variable local conditions. They also conclude that "the shift to a relatively expedient core technology was not consistently related to other technological innovations...horticulture, or local environmental conditions" (p. 303). As no correlations were found with respect to these variables, they were rejected as being causal. One correlation was found to be significant. They state: "In each region...the change to expedient core technology was closely correlated with a shift to sedentism and the first documented occupation of permanent nucleated villages" (p. 303).

6. *Sedentism is assigned causal importance in the shift.* Because of the presence of a correlation between the shift to sedentism and the shift to a more expedient technology, it is postulated that there is a causal relationship between the two, and that: "increasingly expedient lithic technology is a logical consequent of decreased residential mobility" (p. 297). Therefore, decreased residential mobility is argued to be causal.

Parry and Kelly note (p. 299):

the choice of expedient over formal core technology involves a tradeoff between the costs of transporting tools and raw materials (which are high for expedient core technology and low for formal or standardized core technology) and the costs of manufacturing and using tools (relatively high for formal, lower for expedient). It would appear that the benefits of portability outweighed the added costs of standardized core technology in a context of high mobility. Once mobility was reduced, however, there was less incentive to expend the effort to produce and maintain formal tools.

Each technology is argued to be better suited to a given mobility situation, and given the change in mobility, a functional response is the logical consequence.

Parry and Kelly have created a convincing, even classic processual argument.

A SELECTIONIST PERSPECTIVE ON TECHNOLOGICAL CHANGE

The following presents the structure of our selectionist arguments regarding the change documented by Parry and Kelly (1987).

Parry and Kelly observe a pattern in the shift from biface to flake production in many areas of North America. Based on their review of the record this conclusion appears entirely accurate (while we do not believe that the dichotomy between biface- and flake-based technology is the most useful analytic structure with which to examine this technological change, for these purposes we accept their analytic units). Subsequently, we conclude that there is an evident change in the replicative success of the technologies; biface technologies decline in relative frequency with time, flake technologies increase. Selectionist theory dictates that we must consider changing replicative success as being the product of either the operation of stochastic processes, selection, or sorting. We must therefore consider each in turn.

Stochastic processes are rejected for the simple reason that the change in all five regions has the same clear directionality. In addition, we accept Parry and Kelly's arguments that there are clear costs and benefits connected with each technology, suggesting that functional distinctions exist.

As the change is directional, an argument for selection can be made. As population increases in each region either concurrently with or subsequent to the shift in technology, it therefore appears that increased fitness is directly correlated with the technological change, suggesting that the shift in technology may constitute an adaptive change. To complete the argument, it is necessary to identify the selective agent(s) involved, i.e., the reason(s) why one technology was favored over the other in a given selective environment.

Parry and Kelly suggest decreased residential mobility as *causing* the shift. Can decreased residential mobility constitute a selective agent? Not within the selectionist framework, as theory dictates that *evolutionary* explanations lie external to human behavior. We cannot provide evolutionary explanations of behaviors or their products (e.g., changes in the relative frequency of artifact types) *only* in terms of other behaviors (e.g., reduced mobility) as these relationships/correlations are proximate (*sensu* Mayr 1982) at best.

Changes in mobility are only one of a suite of human behaviors that are the products of the action of selective agents. There may be a proximate (functional) relationship between reduced mobility and flake technology, but, if so, changes in mobility are in turn logically a consequence of the operation of selective agents. In general, directional changes correlated among variables may reflect the operation of selective agents (resulting in an evolutionary explanation), a functional or proximate relationship, or the correlation may be merely an association determined by a common cause.

Selective agents are by definition environmental, and the selective equation changes when either the environment changes *or* when new variants are introduced into that environment. The latter precludes charges of environmental determinism being made against the selectionist framework; people generate variants independent of the operation of natural selection and ignorant of future changes in the selective environment (Rindos 1989).

As stated above, an argument for selection can be made because of the clear directionality of the change. Having rejected Parry and Kelly's cause as a selective agent, our next step is to provide one. We do so in the context of our examination of our third alternative, sorting.

It is necessary to establish whether or not sorting has occurred. That is, is the change in technology that we wish to explain linked to other technologies or human behaviors that are being selected? Here we believe that one observation made by Parry and Kelly is particularly important. Parry and Kelly note "...the shift does seem to correlate with the first emphasis on maize as a major staple in the diet of each area" (p. 297).

From a selectionist perspective, we argue that the reduction in mobility *and* changes in technology are linked in some manner to the powerful selective forces favoring maize production (see Rindos 1984 for a description of this evolutionary interaction).

We suggest two possibilities for the structure of this relationship. The first is that flake technology is proximately the product of reduced mobility (as described by Parry and Kelly), and reduced mobility the product of selective agents favoring increased maize production (see Leonard 1989 for a discussion of the role of population growth here). The second is that if flake technology is associated in some manner with the mechanics of agricultural production, increases in the proportion of flake technology in the record may be a product of the subsistence shift toward increased agricultural production, because of the increased importance of the technology associated with agriculture, and perhaps because of the decreased importance of technology associated with hunting.

In each, the change in technology, as well as the reduction in mobility documented by Parry and Kelly (1987) are *both* ultimately the products the actions of selective agents that caused the subsistence shift. Flake technology was present in earlier periods, and its proportional representation merely increased as a product of the linkage.

While we believe that the linkage does exist, it does not constitute pure sorting where the linked trait (flake technology) is of no adaptive significance. A reproductive advantage is certainly conferred by the use of flake technology (compared to no flake technology), and it likely constitutes an adaptation. However, the major reproductive consequences in this example are in terms of agriculture. In other words, there is an *adaptive differential* in flake and agricultural technologies, that must be considered. The linkages suggested above allow us to conclude that the changes in lithic technologies were primarily the product of the operation of selective agents on agricultural production, rather than a product of the operation of selective agents on the lithic technologies *per se*.

CONCLUSIONS

We are aware of two other explanations offered for the trend. Sassaman (1992) suggests that the change reflects the increased visibility of women's activities in the archaeological record as a product of the increasing importance of agricultural production in later time periods. Jeske (1992) discusses the change as a "degeneration of lithic technology" (p. 468) during the late prehistoric periods of eastern North America. Using the framework proposed by Torrence (1983), Jeske argues that "the de-emphasis on lithic technology was an adaptive response to increasing demands on a restricted time and energy budget" (p. 468).

From our perspective these do not constitute evolutionary explanations of phenomena that must be explained in evolutionary terms. Neither the increased visibility (or importance) of women's activities (we choose not to evaluate the major assumptions that are necessary for this characterization to hold), nor increasing demands on a restricted time and energy budget (time stress) constitute selective agents for many of the same reasons that mobility is not, as outlined above. It is interesting to observe that the three alternative explanations that we consider here (Parry and Kelly 1987; Jeske 1992; Sassaman 1992) are not actually competing explanations at all, but only *descriptions* of changes in variable states that are not necessarily incompatible (division of labor, time stress, and mobility), and have nothing to do with cause in evolutionary terms.

In terms or our proposed explanation, we recognize that it is incomplete. Regardless, we believe that our application of selectionist theory not only has provided a more complete explanation of the shift from biface- to flake-based technology but also has demonstrated the potential of selectionist thought. That potential exists largely because *selectionist theory has explanatory components that do not exist in contemporary processual thought.* Evolutionary explanations cannot be constructed within the processual framework because processual archaeology lacks the fundamental theoretical components necessary to address issues of change.

Acknowledgments

We would like to thank Herb Maschner, David Meltzer, Heidi Reed, Anastasia Steffen, and Fritz Taylor for their helpful comments on earlier versions of this paper.

REFERENCES

Braun, D.P., 1990, Selection and Evolution in Non-hierarchical Organization, in: *The Evolution of Political Systems: Sociopolitics in Small-Scale Sedentary Societies* (S. Upham, ed.), Cambridge University Press, Cambridge, pp 62–86.

Dunnell, R.C., 1978, Style and Function: A Fundamental Dichotomy, *American Antiquity* 43:192–202.

Dunnell, R.C., 1980, Evolutionary Theory and Archaeology, in: *Advances in Archaeological Method and Theory*, Vol. 3 (M.B. Schiffer, ed.), Academic Press, New York, pp. 35–99.

Dunnell, R.C., 1989, Aspects of the Application of Evolutionary Theory in Archaeology, in: *Archaeological Thought in America* (C.C. Lamberg-Karlovsky, ed.), Cambridge University Press, Cambridge, pp. 35–49.

Dunnell, R.C., and Feathers, J.K., 1991, Late Woodland Manifestations of the Malden Plain, Southeast Missouri, in: *Late Woodland Stability, Transformation, and Variation in the Greater Southeastern United States* (M.S. Nassaney and C.R. Cobb, eds.), Plenum Press, New York.

Flannery, K.V., Marcus, J., and Kowalewski, S.A., 1981, The Preceramic and Formative of the Valley of Oaxaco, in: *Supplement to the Handbook of Middle American Indians*, Vol. 1 (V.R. Bricker and J.A. Sabloff, eds.), University of Texas Press, Austin, pp. 48–93.

Gould, S.J., and Vrba, E.S., 1982, Exaptation—A Missing Term in the Science of Form, *Paleobiology* 8(1):4–15.

Hofman, J.L., and Morrow, C., 1985, Chipped Stone Technologies at Twenhafel: A Multicomponent Site in Southern Illinois, in: *Lithic Resource Procurement: Proceedings from the Second Conference on Prehistoric Chert Exploitation* (S.C. Vehik, ed.), Occasional Paper No. 4, Center for Archaeological Investigations, Southern Illinois University, Carbondale, pp. 165–182.

Jeske, R.J., 1992, Energetic Efficiency and Lithic Technology: An Upper Mississippian Example, *American Antiquity* 57(3):467–481.

Leonard, R.D., 1989, Resource Specialization, Population Growth, and Agricultural Production in the American Southwest, *American Antiquity* 54:491–503.

Leonard, R.D., and Abbott, A.L., 1993, Theoretical Aspects of Subsistence Stress and Cultural Evolution. Paper presented at the conference on Resource Stress, Economic Uncertainty, and Human Response in the Prehistoric Southwest. Santa Fe Institute, Santa Fe, New Mexico.

Leonard, R.D., and Jones, G.T., 1987, Elements of an Inclusive Evolutionary Model for Archaeology, *Journal of Anthropological Archaeology* 6:199–219.

Mayr, E., 1982, *The Growth of Biological Thought: Diversity, Evolution, and Inheritance*, Harvard University Press, Cambridge, MA.

Montet-White, A., 1968, *The Lithic Industries of the Illinois Valley in the Early and Middle Woodland Period*, Anthropological Papers No. 35, Museum of Anthropology, University of Michigan, Ann Arbor.

O'Brien, M., and Holland, T., 1990, Variation, Natural Selection, and the Archaeological Record, in: *Advances in Archaeological Method and Theory*, Vol. 2 (M.B. Schiffer, ed.), University of Arizona Press, Tucson, pp. 31–79.

O'Brien, M., and Holland T., 1992, The Role of Adaptation in Archaeological Explanation, *American Antiquity* 57(1):36–59.

Parry, W.J., 1983, Chipped Stone Tools in Formative Oaxaca, Mexico: Their Procurement, Production, and Use, Unpublished Ph.D. dissertation, Department of Anthropology, University of Michigan, Ann Arbor.

Parry, W.J., and Christenson, A.L., 1986, *Prehistoric Stone Technology on Northern Black Mesa, Arizona*, Occasional Paper No. 12. Center for Archaeological Investigations, Southern Illinois University, Carbondale.

Parry, W.J., and Kelly, R.L., 1987, Expedient Core Technology and Sedentism, in: *The Organization of Core Technology* (J.K. Johnson and C.A. Morrow, eds.), Westview Press, Boulder, pp. 285–304.

Rindos, D., 1984, *The Origins of Agriculture: An Evolutionary Perspective*, Academic Press, New York.

Rindos, D., 1989, Undirected Variation and the Darwinian Explanation of Cultural Change, in: *Advances in Archaeological Method and Theory*, Vol. 1 (M.B. Schiffer, ed.), University of Arizona Press, Tucson, pp. 1–45.

Sassaman, K.E., 1992, Lithic Technology and the Hunter-Gatherer Sexual Division of Labor, *North American Archaeologist* 13(3):249–262.

Torrence, R., 1983, Time Budgeting and Hunter-Gatherer Technology, in: *Hunter-Gatherer Economy in Prehistory: A European Perspective* (B. Bailey, ed.), Cambridge University Press, Cambridge, pp. 11–22.

Chapter 4

Cultural Virus Theory and the Eusocial Pottery Assemblage

BEN R. S. CULLEN[†]

INTRODUCTION

Cultural virus theory (CVT) is one of three distinct positions which are emerging within what has been termed Darwinian Culture Theory (Durham 1990:190, 1991:183–185) and Cultural Selectionism (Rindos 1986:315). The other two bodies of literature are that of the Meme position (Dawkins 1976, 1982, 1989, 1993; Ball 1984; Delius 1989, 1991; Moritz 1990; Heylighen 1992a,b) and that of the Inclusive Phenotype position (Dunnell 1980, 1989; Boyd and Richerson 1985; Leonard and Jones 1987, and see also this volume; Braun 1990; Durham 1990, 1991; O'Brien and Holland 1990, 1992; Cavalli-Sforza 1991; Neff 1992, 1993). The three positions have much in common, and are, of course, cultural phenomena themselves; each approach would advocate a Darwinian explanation for both its own emergence and that of its nearest philosophical relatives. CVT, for example, would characterize the recent proliferation and diversification of neo-Darwinian approaches as a cultural equivalent of the process whereby the first amphibious vertebrates colonized dry land; other phenomena, such as insects, were already there, but often not in direct competition. Selective metaphysics has colonized a new set of cultural niches, where no neo-Darwinian approaches existed before, namely, the cogni-

---

[†] Deceased.

BEN R. S. CULLEN • Sherwood House, St Dogmael's, Cardigan, West Wales SA43 3LF, United Kingdom.

tive landscape of cultural and psychological fields of inquiry. They have spread so rapidly and so widely in modern academia, that, like butter spread over too much bread, the processes of founder effect, quasi-isolation, drift, and adaptive radiation have produced amazing symbolic diversity from very recent common cultural ancestors. In keeping with the principles of punctuated equilibria advanced by Eldredge and Gould (1972), small populations of ideas erratically distributed throughout the individual minds of partially isolated academic communities allow new concepts to become fixed in local traditions more easily. Such diversity is good. It allows the discovery of a wider range of interpretations, creating a pool of cultural selectionist ideas from which a clearer picture can emerge.

Differences between the three cultural selectionisms are outlined in some detail elsewhere (Cullen 1992, 1993, 1994a,b). CVT may be contrasted with the Meme discourse by the fact that it arose in Australian prehistoric archaeology, in an academic context influenced by ideas such as those of Clarke (1968), Clegg (1978, 1981), Dunnell (1980), Flenniken and White (1985), Fletcher (1977, 1989, 1992), Hodder (1982), Renfrew (1982), Rindos (1986), Shanks and Tilley (1987a,b), White and O'Connell (1982), and Wright (1977), rather than in the context of modern evolutionary biology. The artifact-oriented academic world produced when an essentially Cantabrigean tradition was forced to contemplate the distinctive Australian archaeological record formed a fertile symbolic environment for a wide range of biocultural analogies. Where CVT drew on Dawkins (1976), it was, ironically, directly from neo-Darwinian genetics, and not via the Meme idea. While the Dawkins-derived discourse tends to locate most of the "agency" in the self-replicating meme—the "me" in Dawkins's favorite phrase "duplicate me" (Dawkins 1993:18)—CVT locates the primary agency in human (albeit culturally constituted) consciousness and individual action. The conscious agent is then viewed as domesticating or selectively breeding and actively replicating cultural phenomena, hence the heuristic use of the word *replicatee* in preference to *replicator* (Cullen 1992). CVT nonetheless has employed the word *meme* from time to time (Cullen 1990, 1994a:Chapter 10), as a convenient word for cultural hereditary material, while advocating a replicatee-type interpretation of the word.

Differences between CVT and the Inclusive Phenotype position, as distinct from between it and the meme discourse, involve the conception of human individuals as ecological assemblages rather than individual phenotypes, and the conception of human populations as larger ecological assemblages rather than true Darwinian populations, arguments which are developed elsewhere (Cullen 1993, 1994a:Chapter 5). At a more general level, CVT may also be distinguished from other cultural selectionist positions by the fact that it is explicitly aligned with aspects of modern psychobiology, a model of the brain known as *neural Darwinism* (Edelman 1989, 1992; Cullen 1994a:Chapter 9).

Edelman concepts such as "degeneracy" and others are explored as potential starting points for building a neo-Darwinian model of cultural mutation and variation.

Science, Religion, and Cultural Virus Theory

Having briefly outlined a cognitive map of the three forms of Cultural Selectionism, attention will now be focused on the CVT position, although the other two positions will be referred to from time to time by way of contrast. First, a general overview of CVT will be attempted through reference to two of the most ubiquitous kinds of cultural assemblage of the modern world, science and religion. Here the arguments of the meme position have attracted a great deal of public attention (Dawkins 1992, 1993), although the alternative "symbiotic domestication" arguments of CVT have not been ignored (Cullen 1992). The key difference between the two positions is where the primary initiative or agency is best located. Both positions allow cultural phenomena to adopt an essentially parasitic or pernicious relationship to their hosts, but while the meme view favors explanations where human intention is not important, CVT takes the opposite view. As implied above, ideas are modeled as actively selected, propagated, physically replicated, and even genetically engineered by human agents through a variety of cognitive processes, with the result that either symbiotic or parasitic ecological relationships can emerge.

As Rindos has demonstrated so convincingly (Rindos 1980, 1984), the relationships between humans and their domesticates, whether plant or animal, are not, in principle, fundamentally different in Darwinian terms to many other parasitic or symbiotic partnerships in nature, such as that between fungi and algae to form the structures known as lichen. CVT extends this principle into the realm of culture, balancing it against principles of parasitism.

CVT-type parasitic situations emerge through culturally informed subjective assessments of the value of a particular set of ideas, such that the negative effects of maintaining a traditional practice may come to outweigh the positives, or because of the fact that humans may become completely dependent on certain ideas despite their pernicious effects. Meme-type "unintentional" parasitic situations are not excluded from CVT, it is only that they are judged to be marginal. Assessment of the value of a particular idea, practice, or artifact is affected by other cultural phenomena, which is in turn affected by other ideas again, and so on in an endless chain of signifiers. As Derrida has shown, ambiguities are endemic in symbolic frameworks (Miller 1993:119); any value is thereby distributed throughout many structures, and it becomes all but impossible to distinguish parasitic ideas from symbiotic ones.

By locating the *primary* agency in human consciousness (there is still some space left for material culture to be conceived as active [Hodder 1982], not passive), CVT thereby casts *all* cultural phenomena in the role of viral phenomena, the vast majority of which would be considered "domesticates" if they were animals or plants. This is another way of saying that a population of human individuals, together with their rituals and artifacts, must be considered as an ecological assemblage of genetically transmitted and culturally transmitted phenomena, rather than as a single population of organisms or phenotypes. Given the genealogical independence of artifacts and ideas, they are *ecologically equivalent* to domesticates; yet being unable to sustain most of the processes associated with "life" (such as breathing, growing, and self-reproducing) to the same extent as organisms, some other term is required. The term v*iral phenomena* was chosen to capture this paradoxical combination of dependence and independence, and of living and nonliving characteristics.

So, while Dawkins's meme position (1992, 1993) would characterize religion as a disease of the mind, CVT would offer a somewhat different account, characterizing religions as vast ecological assemblages of both symbiotic and parasitic ideas, heuristic metaphors threaded together in a cocktail of truth and ambiguity. Priests are the custodians of these ideas, in the same way that a shepherd might look after a flock of sheep, or an apiarist a hive of bees. Such assemblages will, of course, fall prey to parasitic phenomena themselves. But it is no simple matter to look at any one idea and say "this is a disease," simply because it may appear to be demonstrably untrue. This is because an arbitrary article of faith may in fact support a series of other symbolic structures which are either relatively truthful, or are very beneficial falsehoods, as the ecological equivalents of scavengers in the food web of the savannah. Tearing out "wrong" ideas may cause the delicate symbolic ecology of a religion to come tumbling down, leaving an uninhabitable cultural desert in its place.

Stories from the life of Christ, for example, may be ambiguous or metaphorical in their particulars, but a general celebration of his life might contribute to adherence to aspects of Christian philosophy, such as "love thy neighbor" or generosity to others, principles which might actually benefit an individual or community in certain contexts. Alternatively, faith in the idea that Christ was "The Son of God," a statement which would appear vulnerable to deconstruction, could merely express the idea that *all* people are the children of God in Christian ideology, or serve to ensure that a great philosopher is held in sufficient respect by those who might otherwise be unable to grasp the significance of his philosophical achievements.

Dawkins views science in a rather different light than religion, and does not go to any effort to characterize it as a parasite or disease; and here the meme position and CVT are again very different. CVT would insist that scientific discourse involves much of the same kind of ecological assemblage of symbiotic and parasitic ideas as religion. Scientists may be described as the priests or shamans of a religion based around doubt, just as many other religions are based around faith. Science cannot be characterized as some kind of wholly objective enterprise which is qualitatively different from religion, or as an activity where absolute "truth" is the only means by which a scientific idea can become established. Metaphor, rhetoric, personal charisma, and many other subjective factors may also play their part.

Knight (1991:7–9), for example, has discussed how fallacious "good of the species" arguments in biology may be linked to social-democratic and corporatist political sentiments, and how the rise of both sociobiology and Marxism can be linked to reactions to this sociopolitical milieu, incorporating basic self-interest into frameworks of individualism and socialism, respectively. Moreover, Miller has described how Derrida has sought to unmask the rhetorical dimensions of rationalist writings:

> whether reading Plato or Rousseau, Derrida probed for inconsistencies, trying to unmask hypocrisies that were symptomatic of the ambiguities inherent in language and thought...in classical rationalists, by contrast, he looked for references, damning because disavowed, to imagination and metaphor. (Miller 1993:119)

More recently, Golinski has discussed the work of Biagioli (1993), which situates the scientific ideas of Galileo in the context of rhetorical honor-duels at the Florentine court of the Medicis (Golinski 1993:22). And it is easy to overlook the role which the skillful rhetoric of early Darwinians played in the rise of selectionist metaphysics in biology. Consider, for example, how often the famous words of Darwin's friend Joseph Hooker, who remarked that he would rather have an ape for a grandfather than a man who would introduce ridicule into a serious scientific debate (Desmond and Moore 1992:497), are referred to in modern evolutionary biology. Of course, the *presence* of *metaphor* in scientific discourse does not prove that there is an *absence* of *objectivity*. It is just that scientific and religious persuasion both involve the strategic combination of "truth" and rhetoric, although the relative contribution of each can vary greatly.

ARTIFACTS AS VIRAL PHENOMENA

For Dawkins, the virus analogy is used pejoratively; no attempt is made to apply it to all of culture, all knowledge, or all ideas, as has already been shown with respect to his approach to science (Dawkins 1993). In CVT, on the other hand, a neutral category of "viral phenomena" is used, which includes all ideas, behavior, and artifacts. This difference is a result of the fact that CVT applies the notion of a viral phenomenon to culture according to a very detailed and specific set of principles, while Dawkins's use of the word *virus* is really only employed as a general metaphor for the notion of a parasite.

In other words, Dawkins's "virus" is a direct and unamended analogy between DNA and culture; an assertion of symmetry between two self-replicating entities, memes and genes. Other words such as *louse, flea, plasmid,* or *bacterium* would capture the Dawkins notion of the self-replicating meme just as accurately. No appeal is being made to anything which is peculiar to viruses and viruses alone. Some evidence for this can be found in the fact that Dawkins apparently makes no attempt to distinguish between DNA viruses, which are true self-replicators, and RNA retroviruses, which contain only RNA and are therefore not able to replicate themselves quite so directly or autonomously.

CVT, by contrast, employs the notion of "viral phenomena" to capture similarities *and differences* between DNA and culture, qualities which distinguish ideas and artifacts from self-replicators, characterizing them as structures which find their *nearest* genetic equivalents in RNA retroviruses. It is important to stress the word *nearest*, as the category is not intended to deny artifacts or ideas any qualities which RNA retroviruses might turn out not to possess, or vice versa. The notion of the cultural viral phenomenon is merely intended to provide a heuristically superior "middle ground" alternative to the "cultural replicators" of the meme discourse on the one hand, and the "cultural traits" of the Inclusive phenotype position on the other (Cullen 1993). It is a title of ambiguity, and thereby of precision, since cultural phenomena are themselves ambiguous.

What, then, are the constituent principles of the notion of viral phenomena? The reader is reminded that the term *viral phenomena* is an artificial generic category designed to summarize as many of the peculiar qualities of cultural phenomena as possible (Cullen 1990:63, 1993:198), not to limit them to a role as gigantic double-gangers of microbial viruses. It is an attempt to place ideas, rituals, and artifacts on a cognitive map of biological phenomena, without denying them any of their unique qualities:

1. *Genealogical independence.* Firstly, and fundamentally, a viral phenomenon displays the potential for genealogical independence from its host, such that host and guest reproduce at different times and by different means. This principle is derived directly from evolutionary biology (Dawkins 1982, 1989:234–267), not via the meme discourse. All cultural phenomena fulfill this proviso by definition, and consequently even "good memes" cannot be considered as human traits (contrary to Dawkins 1993), but must be considered a class of symbiotes, since unlike good microbial viral genes they cannot integrate into the host genotype.
2. *Dependence.* Despite its genealogical independence, a viral phenomenon is very dependent on the host body. It displays periods of inti-

mate integration when not undergoing social transmission. Like an RNA retrovirus which has reverse transcribed (Kuby 1992:463–466; Eddy and Walden 1993:32; Greene 1993:69), the hereditary material of cultural phenomena is physically intertwined, even at times synonymous, with parts of the host.

3. *Intrinsic neo-Darwinian logistics.* The evolution of viral phenomena is neo-Darwinian in every particular, despite the fact that they may be "acquired," as evolved objects, by the host, in a "Lamarckian" manner. When we acquire a virus, we do not acquire a new trait, but rather become associated with an object which brings with it its own phylogenetic heritage and selective history. For this reason it is possible to classify microbial entities phylogenetically, according to patterns of descent, independently of host phylogeny (Fenner et al. 1974:1–33; Postgate 1989; and despite bacterial virion exchange as described by Sonea 1988). By implication, artifacts must be conceived as being fundamentally constrained by cultural traditions in a perfectly neo-Darwinian manner; i.e., that there are families of artifacts, just as there are families of languages in linguistics (Ruhlen 1991), but that artifact families have the potential to be independent from human families.

4. *Limited intrinsic agency.* A viral phenomenon does not think, plan, or feel; it does not have its own nervous system. It cannot move or behave to the extent of a more complex organism, although its presence may influence the behavior of such organisms, and it is possible to argue that they are "living" (Slap 1991). Similarly, ideas, rituals, and artifacts have no intrinsic consciousness, although they may awaken states of consciousness in their hosts, and would seem to be (at least) closer to living things than natural inorganic phenomena such as glaciers or boulders.

5. *Capacity for physical separation of genotype and phenotype.* Another factor which makes viruses more like cultural phenomena than any other kind of structure is the fact that it is possible for a viral phenotype and its hereditary material to exist independently of each other, which happens when a virion attaches itself to a host cell and injects its hereditary material into the cell, with the envelope remaining outside (Greene 1993:69). Similarly, cultural phenotypes (such as rituals and artifacts) may be conceptually distinguished from the knowledge of their performance or manufacture which exists in human minds.

6. *An increased dependence on host agency during the act of replication.* Here cultural phenomena clearly fit the principle more closely still than RNA retroviruses, which carry bits of their phenotypes with

them into the host cell, enzymes such as reverse transcriptase (Greene 1993:68) capable of initiating translation into DNA. Although it may be possible to view artifacts and rituals as containing clues about how they may be manufactured or reproduced, which may "initiate cultural reverse transcription" as it were, it would seem that this capacity is not quite as sophisticated as in RNA- or DNA-based viral phenomena.

7. *One-, two-, or three-tier structure.* Cultural viral phenomena have three basic tiers to their structure. The first level is that of the idea, remembered action sequence or manufacturing concept, which consists of a pattern of strengthened synapses distributed within a brain (Delius 1989, 1991; Edelman 1989, 1992). Such structures constitute the hereditary material of cultural phenomena. The second level is that of action and ritual, which consists of instances of behavior, or of the regulation of such behavior by other "coordinating" ideas. The third level is that of the material consequences of such behavior, that of the artifact, a dimension which some cultural phenomena may not necessarily possess.

8. *A continuous spectrum of adaptive strategy.* As a emergent property of principle 1, viral phenomena, despite their dependence on their hosts for so many facilities, can vary all the way from being perfectly compatible with their hosts, to being dramatically exploitative (Mitchison 1993:105). Cultural phenomena seem to display a comparable spectrum of variation, ranging from survival-promoting artifacts such as kayaks and igloos, all the way to lethal systemic drugs. This is in contrast to a true "trait" or organ, which has only one polarity in its adaptive strategy, that of optimum benefit to the rest of the phenotype, and not a continuous spectrum of possible adapted states.

9. *People are ecological assemblages of cultural and genetically emergent phenomena.* People and populations thereof are not excluded from CVT; rather, they are promoted to the level of ecological assemblages, which evolve differently than single organisms. For example, the interests of constituent structures of an ecological assemblage are not synonymous, whereas the interests of constituent organs of one organism are. The genetically emergent dimensions of the human body function as the domesticator, while cultural phenomena fulfill the role of domesticates. Since the domesticator actually constitutes its own identity from ideas, actions, rituals, and artifacts, this relationship is fundamentally ambiguous. The human mind, more than that of any other animal, is an ongoing theater of Darwinian schizophrenia. It is a multiplicity of many genotypes and phenotypes, not a uni-

ty of one. The genetically emergent dimension of the human individual could be imagined as a kind of local deity which surrounds and pervades (embodying *parts* of itself in terms of) a personal pastiche of cultural phenomena.

This summary of the notion of viral phenomena provides a suite of principles to manipulate in the explanation of regional sequences of any kind of artifact. There are several more explanatory components to CVT, such as the idea of artifact eusociality, but it will be more productive to present these additional principles in concert with those mentioned above in the context of some familiar archetypes of the archaeological record.

THE EUSOCIAL POTTERY ASSEMBLAGE

CVT, like any selective theory, is all about explaining the results of archaeological excavations after they have been described and published as archaeological reports. Where cultural selectionist explanatory structures are more established in the discipline, they would be expected to have more influence on decisions about excavation and postexcavational analysis. Where they are not so established, an archaeologist is more likely to require an understanding of cultural selectionism only after questions of what, where, and when have been at least provisionally answered, and attention is being turned to how and why. CVT is designed to act as a provocative set of answers to the question of "what?" in archaeology, one which is intended to suggest a framework to questions of "how?" and "why?"

The archaeological implications of CVT may be understood in terms of a fundamental consideration of the question "What is an artifact?" and the related question "What is an artifact assemblage?" Below, this question is considered with reference to the generically more specific questions "What is a pot?" and "What is a pottery assemblage?" A range of archaeological and anthropological implications of CVT have been discussed elsewhere (Cullen 1993, 1994a,b), but the main theme outlined here is the idea that most artifacts, when conceived as viral phenomena, would seem to exhibit an extraordinary reproductive structure which is considered to be rare and challenging in modern evolutionary biology, and which is quite unknown in almost all of the world's DNA- or RNA-based species, including microbial viruses. If this remarkable phenomenon was ever discovered in the biology of viruses, it would probably make an incredible impact, not only on virology, but on evolutionary biology as a whole. Yet, disguised as this phenomenon is in the familiar patterns of everyday archaeological practice and domestic life, it escapes the notice of archaeologists. This is the phenomenon known as "eusociality," of which the most well-known examples are those of social insects such as bees and termites, where a range of sterile, functional "worker" individuals are produced in order to serve the survival of a minority of reproductive individuals which act as "breeders." Such concepts have even been exploited by the cinema in box office hits such as the *Alien* series, and *Arachnophobia*, whose fantasy life-forms have eusocial colony structures, although the average cinema goer would not be familiar enough with modern evolutionary biology to notice. And if this particular CVT-generated hypothesis is correct, the archaeological record is littered with artifact assemblages, the hereditary structure of which is every bit as complicated as a beehive or termite nest. Cultural fields of inquiry could well be sitting inadvertently on a discovery of huge significance to the life sciences, a virtual kingdom of cultural phenomena where eusociality is as common as it is rare in DNA-based phenomena. This could also explain the long delay in the Darwinization of cultural fields of inquiry, since the remarkable altruism of sterile bees, ants, and termites was also a major conundrum for neo-Darwinism, and one which took many years to solve.

First, it is pertinent to communicate some of the fascination that eusocial structures such as beehives hold for many biologists. Here is a short description:

> A social insect colony is a huge family, usually all descended from the same mother. The workers, who seldom or never reproduce themselves, are divided into a number of distinct castes, including small workers, large workers, soldiers and highly specialized castes like honey pots (individuals with swollen abdomens which act as food storage vessels). Reproductive females are called queens. Reproductive males are sometimes called drones or kings. In the more advanced societies, the reproductives never work at anything except procreation, but at this one task they are extremely good. They rely on the workers for their food and protection, and the workers are also responsible for looking after the brood. In some ant and termite species the queen has swollen into a gigantic egg factory.... (Dawkins 1989:172)

For a long time this kind of social structure was thought to occur only in social insects, but recently it was discovered in a hairless and virtually blind rodent, the naked mole rat, which is found in arid areas of Kenya, Somalia, and Ethiopia (Dawkins 1989:173, 313). This mammal has a main lineage of "queen" and "drone" reproductive forms, and a continuum of sterile forms which spend their lives eating, sleeping and carrying out the various activities required to maintain the colony.

Naked mole rats inspired so much interest among biologists that captive colonies made of many meters of labyrinthine transparent tubing were set up in at least two universities and at the London Zoo (Dawkins 1989:314) within a relatively short time, attracting the attention of such notables as Jennifer Jarvis, Richard Alexander, and Paul Sherman, in addition to that of Dawkins, while Robert Brett undertook field observations of mole rat colonies in Kenya. In some sense, these insect and mole rat colonies are multiorganismic equivalents of single organisms, with each nonreproductive caste fulfilling a role similar to that of organs such as the liver or the skin; these structures cannot reproduce themselves, and can only work toward the general reproductive success of testes and ovaries, organs which specialize in producing the next generation. But in social insects and naked mole rats the principle of altruistic sterility has been shifted to the level of whole bodies, where the modular unit of adaptive differentiation and specialization becomes the organism rather than the cell. The benefit for the social insect colony is that now there are a multiplicity of structures, a multiplicity of animated organs if you like, each capable of moving, living, and dying independently of each other, which can serve the reproductive interests of the main lineage in a very sophisticated manner. Individual cells are far more limited in their movements than are individual organisms.

I would like to submit the proposal (see also Cullen 1990:66, 1994a: Chapter 10) that this kind of "multiobject" reproductive strategy, where sterile nonreproductive structures are produced in order to facilitate the reproductive success of a central lineage of reproductive forms, is also present in cultural fields of inquiry. Indeed, I would like to go further than this, and tentatively submit the possibility that such complicated reproductive structures are not only *present* in cultural phenomena, but are in fact the *norm*. For instance, imagine the various artifacts present in the modern home environment, such as computers, furniture, crockery, cutlery, books, stereos, and so on. How many of such objects are actually produced in the home? And how many of such objects will ever be *reproduced* where they are found? Human individuals surround themselves with a plethora of sterile objects which can neither reproduce themselves, nor will they *be reproduced* by their owners. Surely the key to the explanation of such phenomena lies in the way they divert resources back to the industries which reproduce them. Like worker bees, their role is to obtain energy from their environments, through the symbolic resonance they excite in the minds of the people who bought them and might buy again, or in the minds of visitors to the household. This symbolic resonance is not so much divided into distinct categories of functional and aesthetic, but is rather a cocktail of perceived beauty, perceived functional efficacy, and any number of additional symbolic dimensions. The stronger the psychological profile a worker artifact excites, the greater the desire to buy or trade for such phenomena, and the more energy is diverted back into the industry. This energy is then used both to replicate the manufacturing industry and to produce more nonreproductive worker objects to go out and forage in new human communities.

There is only enough space to consider the form that this kind of replicative strategy might take in one kind of artifact in even a cursory manner, and ceramic artifacts have been chosen. First, let us imagine what the hereditary material of pottery assemblages is made of. This material consists of the knowledge of everything to do with making ceramic artifacts: knowledge of where to look for raw materials, of extraction, preparation, and mixing of such materials, and of the manufacture of the pots themselves, of hand techniques such as coiling and the manufacture and use of pottery wheels and kilns, and lastly of the drying, firing, and marketing of the finished pots. This is the hereditary material of the pottery assemblage, and it is housed in the minds of one or more potters, just as the genome of a beehive or mole rat colony is housed in the bodies of reproductives like queens and drones. Except, of course, for the fact that the body that the pottery hereditary material inhabits is not a body of its own making, nor wholly of its own using. In this sense it is like a hive which is built into a living structure, such as a tree in a rain forest, which is inhabited and used by many other phenomena all seeking their own reproductive success, not least the reproductive organs of the tree.

All of the behavior patterns and manufacturing techniques performed by the potters, along with the pots themselves, constitute the phenotype of the above-mentioned hereditary material, as the second and third tiers of the pottery industry. Together they constitute the body of the viral phenomenon we call a pottery assemblage; which is best envisaged as a whole colony of viral phenomena, a multiobject assemblage in which different kinds of structure specialize at different tasks, many of which are not directly concerned with reproduction. As far as the potter is concerned, the pottery-making knowledge is gradually acquired over a period of apprenticeship. In some sense, from time to time in his or her life, the potter may be thought of in sociobiological terms, as an individual attempting to accumulate and control the flow of resources in his or her community, in such a way as to provide for a family of genetic relatives, despite increasing interference from a personal assemblage of acquired ideas.

As already discussed in more general terms, the relationship between potter and pot may be viewed as ecologically equivalent to the relationship between shepherd and sheep. We have a potentially symbiotic relationship, which is maintained through the discharge of various forms of human agency and initiative. Yet once the pottery knowledge has been acquired it can now influence the thought processes to various degrees, depending on how important being a "potter" becomes to the identity of a pot maker. This is of course thematically similar to the degree to which the role of being a shepherd colonizes the consciousness (and unconsciousness) of someone who becomes involved with herding sheep. While parasitic relationships may emerge, where potters become completely celibate and spend their entire lives making pots and contributing nothing to their family, more often the Darwinian stakes involve *relative* amounts of benefit in an ongoing symbiotic relationship. It is not a question of the pottery industry getting all of the potter's energy for nothing, but of getting *more* benefit than the potter's family, in a situation where both will usually benefit to some degree.

For example, we have the structures which inhabit the minds of the potters, the manufacturing knowledge itself. These are the "queen" structures, as it were, and consist of a series of neuronal groups distributed throughout the potter's brain, which display a particular pattern of connectivity which allowed them to represent pottery-making techniques better than any others, according to the mechanisms of neuronal group selection developed by Edelman (1989, 1992). While the potter was an apprentice, these particular patterns of neuronal connectivity were "selected" and amplified during the process of instruction of the apprentice by another tradesman. The ultimate "goal" of this assemblage of pottery-making knowledge (although, of course, it has no intrinsic consciousness) is to replicate itself, and to draw new apprentices into the pottery trade.

The role of the pots is then ecologically analogous to that of worker bees, termites, or naked mole rats. Obviously, pots cannot physically collect energy and bring it back to the potter's workshop. Their service to the pottery industry is far more subtle than that, and does not involve active behavior but rather an emerging involvement in the activities of the various human agencies and intentions which surround them, and which they sometimes embody. Some worker pots will be put to use in the workshop, facilitating the daily practice of the production of more worker pots, or even acting as archetypes for imitation by apprentices. These are equivalent to the nurses in a beehive. Other pots will be manufactured for local markets, diverting local products and currencies back to the potter's workshop. Their success in this task will depend on how well they are *perceived* to hold food or drink, act as storage or vessels, how good they look, what social information they can convey and so on; that is, they are successful to the extent that they fit the culturally and individually specific notions of efficacy and beauty which exist in the minds of the people of the local community. Such pots fulfill an equivalent role to foraging workers in the hive, except that they make only one journey, and all movements of resources occur as a result of human agency. And finally, in a community which is visited by traders, a remarkable dimension of the pottery phenotype may emerge. Ceramic artifacts which are traded can extend the catchment area of the pottery assemblage for many thousands of miles, as in the case of the trading of fine porcelain. Such large scale movements can only serve the survival of the pottery trade if a significant percentage of the traded price eventually makes it back to the original workshop, either in the form of disposable goods or currency which can be channeled into the workshop, or in the form of eager foreign apprentices wanting to learn to make fine porcelain themselves.

While there is no space to expand on this purely hypothetical model of pottery assemblages in the archaeological record, it is hoped that the brief sketch has helped the reader to envisage one of the possible implications of CVT. An actual case study would be impossible in a paper of this size, particularly in the case of a new theoretical framework, which has to be at least summarized before any of the archaeological implications can be explored.

CONCLUSION

Attention in the last section of the paper has been focused on just one of the basic implications of CVT, and it is hoped that readers will be prepared to suspend disbelief until other aspects of the theory can be made available. The aspect of CVT outlined here has been the remarkable potential of artifacts to form extremely complicated eusocial object assemblages which are selected in history according to their ability to arouse the attention of people in such a way that a central core of technological knowledge is reproduced. Presumably, such artifact eusociality reaches its crescendo in the form of widely traded items, but the principle could be applied just as easily to less organized manufacturing industries.

How does the eusocial hereditary structure of pottery assemblages fit in with the general CVT notion that cultural reproduced phenomena are ecologically equivalent to domesticates? Does this mean that artifacts are only analogous to domesticated eusocial insects, such as bees? The answer is most definitely no. Once the vision of a eusocial pottery or stone artifact assemblage is firmly grasped, it quickly becomes apparent that not only are we surrounded by other eusocial artifact assemblages, such as computers, sheets of paper, household crockery and cutlery, stoves, automobiles, garden tools, and packaged foods, but almost as many eusocial animal and plant structures too. These would include cut flowers, desexed family pets, meat-producing animal populations which supply human communities with nonreproductive individuals such as suckling pigs and fat lambs, castrated draught animals such as oxen, sterile hybrid varieties of animals and plants, and the list could be extended indefinitely. Each one of these organisms has little chance of reproducing themselves, and must presumably function as sterile workers for other copies of their genes. Thus, it can be seen that by exploring only one angle of CVT, we are offered not only a new vision of the nature of the archaeological record, but new visions of the nature of genetic process as well. Human agents seem to be able to impose an artificial eusocial structure on domesticated populations such that the vast majority of a particular variety of animal or plant are sacrificed in various ways which nonetheless allow the dramatic reproductive success of a selected few.

If even a few of the above cases of eusocial artifact assemblages are correct, archaeology has the potential to turn the tables on modern evolutionary biology for the first time, and actually submit extraordinary cases of neo-Darwinian dynamics which biologists have never encountered. It is important to note that this is not to say that culture is any less *neo-Darwinian* in its logistics, as some have remarked; the uniqueness of cultural phenomena does not consist of Lamarckian processes. It is rather that we find unique but perfectly neo-Darwinian inheritance structures in artifacts, a plethora of eusocial artifact assemblages which could well be responsible for much of the distinctive ontology of the archaeological record.

REFERENCES

Ball, J.A., 1984, Memes as Replicators, *Ethology and Sociobiology* 5:145–161.
Biagioli, M., 1993, *Galileo, Courtier: The Practice of Science in the Culture of Absolutism*, University of Chicago Press, Chicago.
Boyd, R., and Richerson, P.J., 1985, *Culture and the Evolutionary Process*, University of Chicago Press, Chicago.
Braun, D.P., 1990, Selection and Evolution in Nonhierarchical Organization, in: *The Evolution of Political Systems* (S. Upham, ed.), Cambridge University Press, Cambridge, pp. 63–86.
Cavalli-Sforza, L.L., 1991, Genes, People and Languages, *Scientific American* 265:72–78.
Clarke, D.L., 1968, *Analytical Archaeology*, 2nd ed. (B. Chapman, ed.), Methuen, London.
Clegg, J.K., 1978, *Mathesis Words, Mathesis Pictures*, Unpublished MA thesis, Department of Anthropology, University of Sydney.
Clegg, J.K., 1981, *Notes Toward Mathesis Art*, Clegg Calenders, Sydney.
Cullen, B.R.S., 1990, Darwinian Views of History: Betzig's Virile Psychopath versus the Cultural Virus, *Crosscurrents* 4:61–68.
Cullen, B.R.S., 1992, Cultural Viruses and Darwin's Fools, *The Independent* May 9, 1992:33.
Cullen, B.R.S., 1993, The Darwinian Resurgence and the Cultural Virus Critique, *Cambridge Archaeological Journal* 3:179–202.
Cullen, B.R.S.,1994a, *The Cultural Virus*, Unpublished Ph.D. Thesis, Department of Prehistoric and Historical Archaeology, University of Sydney.
Cullen, B.R.S., 1994b, Social Interaction and Viral Phenomena, in: *Power, Sex and Tradition: The Archaeology of Human Ancestry* (S. Shennan and J. Steele, eds.), Routledge, London, pp. 1–16.
Dawkins, R., 1976, *The Selfish Gene*, Freeman, San Francisco.
Dawkins, R., 1982, *The Extended Phenotype*, Freeman, San Francisco.
Dawkins, R., 1989, *The Selfish Gene: New Edition*, Oxford University Press, Oxford.
Dawkins, R., 1992, The Mind is Prey to Virus and Parasite. *The Independent* May 2, 1992:30.
Dawkins, R., 1993, Is Religion Just a Disease? *The Daily Telegraph* December 15, 1993:18.
Delius, J.D., 1989, Of Mind Memes and Brain Bugs: A Natural History of Culture, in: *The Nature of Culture* (W.A. Koch, ed.), Studienverlag Brockmeyer, Bochum, pp. 26–79.
Delius, J.D., 1991, The Nature of Culture, in: *The Tinbergen Legacy* (M.S. Dawkins, T.R. Halliday, and R. Dawkins, eds.), Chapman and Hall, London, pp. 75–99.
Desmond, A., and Moore, J., 1992, *Darwin*, Penguin, London.
Dunnell, R.C., 1980, Evolutionary Theory and Archaeology, in: *Advances in Archaeological Method and Theory*, Vol. 3 (M.B. Schiffer, ed.), Academic Press, New York, pp. 38–99.
Dunnell, R.C., 1989, Aspects of the Application of Evolutionary Theory in Archaeology, in: *Archaeological Thought in America* (C.C. Lamberg-Karlovsky, ed.), Cambridge University Press, Cambridge, pp. 35–49.
Durham, W.H., 1990, Advances in Evolutionary Culture Theory, *Annual Review of Anthropology* 19:187–210.
Durham, W.H., 1991, *Coevolution: Genes, Culture, and Human Diversity*, Stanford University Press, Stanford.
Eddy, P., and Walden, S., 1993, AIDS—A Special Investigation, *Telegraph Magazine* November 20, 1993:28–36,57–62.

Edelman, G.M., 1989, *Neural Darwinism: The Theory of Neuronal Group Selection*, Oxford University Press, Oxford.
Edelman, G.M., 1992, *Bright Air, Brilliant Fire*, Allen Lane, Penguin, London.
Eldredge, N., and Gould, S.J., 1972, Punctuated Equilibrium: An Alternative to Phyletic Gradualism, in: *Models in Palaeobiology* (T.J.M. Schopf, ed.), Cooper and Company, San Francisco, pp. 82–115.
Fenner, F., McAuslan, B.R., Mims, C.A., Sambrook, J., and White, D.O., 1974, *The Biology of Animal Viruses*, 2nd ed., Academic Press, New York.
Flenniken, J.J., and White, J.P., 1985, Australian Flaked Stone Tools: A Technological Perspective, *Records of the Australian Museum* 36:131–151.
Fletcher, R.J., 1977, Settlement Studies: Micro and Semi-micro, in: *Spatial Archaeology* (D.L. Clarke, ed.), Academic Press, New York, pp. 47–162.
Fletcher, R.J., 1989, The Messages of Material Behavior: A Preliminary Discussion of Non-verbal Meaning, in: *The Meanings of Things: Material Culture and Symbolic Expression* (I. Hodder, ed.), Unwin Hyman, London, pp. 33–40.
Fletcher, R.J., 1992, Time Perspectivism, Annales, and the Potential of Archaeology, in: *Archaeology, Annales, and Ethnohistory* (A.B. Knapp, ed.), Cambridge University Press, Cambridge, pp. 35–49.
Golinski, J., 1993, Shining Star Who Courted Controversy, *The Times Higher Education Supplement* December 17, 1993:22–23.
Greene, W.C., 1993, AIDS and The Immune System, *Scientific American* 269:66–73.
Heylighen, F., 1992a, "Selfish" Memes and the Evolution of Co-operation, *Journal of Ideas* 2:77–84.
Heylighen, F., 1992b, Non-rational Cognitive Processes as Changes of Distinctions, in: *New Perspectives on Cybernetics, Self-Organization, Autonomy and Connectionism* (R. Trappl, ed.), World Science, Singapore, pp. 3–10.
Hodder, I., 1982, *Symbols in Action*, Cambridge University Press, Cambridge.
Knight, C., 1991, *Blood Relations: Menstruation and the Origins of Culture*, Yale University Press, New Haven.
Kuby, J., 1992, *Immunology*, Freeman, San Francisco.
Leonard, R.P., and Jones, G.T., 1987, Elements of an Inclusive Evolutionary Model for Archaeology, *Journal of Anthropological Archaeology* 6:199–219.
Miller, J., 1993, *The Passion of Michel Foucault*, HarperCollins, London.
Mitchison, A., 1993, Will We Survive? *Scientific American* 269:102–108.
Moritz, E., 1990, Memetic Science: I—General Introduction, *Journal of Ideas* 1:1–23.
Neff, H., 1992, Ceramics and Evolution, in: *Advances in Archaeological Method and Theory*, Vol. 4 (M.B. Schiffer, ed.), University of Arizona Press, Tucson, pp. 48–63.
Neff, H., 1993, Theory, Sampling, and Analytical Techniques in the Archaeological Study of Prehistoric Ceramics, *American Antiquity* 58:23–44.
O'Brien, M.J., and Holland, T.D., 1990, Variation, Selection, and the Archaeological Record, in: *Advances in Archaeological Method and Theory*, Vol. 2 (M.B. Schiffer, ed.), University of Arizona Press, Tucson, pp. 177–197.
O'Brien, M.J., and Holland, T.D., 1992, The Role of Adaptation in Archaeological Explanation, *American Antiquity* 57:36–59.
Postgate, J., 1989, Microbial Happy Families, *New Scientist* January 21, 1989:40–44.
Renfrew, C., 1982, *Towards an Archaeology of Mind*, Cambridge University Press, Cambridge.
Rindos, D., 1980, Symbiosis, Instability, and the Origin and Spread of Agricultural Systems: A New Model, *Current Anthropology* 21:751–772.
Rindos, D., 1984, *The Origins of Agriculture*, Academic Press, New York.
Rindos, D., 1986, The Evolution of the Capacity for Culture: Sociobiology, Structuralism and Cultural Selectionism, *Current Anthropology* 27:315–331.

Ruhlen, M., 1991, *A Guide to the World's Languages*, Vol. 1, Hodder and Stoughton, Melbourne.
Shanks, M., and Tilley, C., 1987a, *Reconstructing Archaeology: Theory and Practice*, Cambridge University Press, Cambridge.
Shanks, M., and Tilley, C., 1987b, *Social Theory and Archaeology*, Polity Press, Cambridge.
Slap, J.K., 1991, Virile Viruses, *Australian Natural History* 23:668.
Sonea, S., 1988, The Global Organism, *The Sciences* 28:38–56.
White, J.P., and O'Connell, J.F., 1982, *A Prehistory of Australia, New Guinea and Sahul*, Academic Press, New York.
Wright, R.V.S., 1977, Introduction and two studies, in: *Stone Tools as Cultural Markers: Change, Evolution, Complexity*. Canberra: Australian Institute of Aboriginal Studies, Canberra, pp. 1–4.

Chapter 5

Organized Dissonance
Multiple Code Structures in the Replication of Human Culture

ROLAND FLETCHER

INTRODUCTION

Since culture is plainly not the same as biology, its replicative system should be unlike that of genetics in some profound way. The central issue of a cultural equivalent of the Darwinian model of evolution is that the nature and operation of the cultural equivalent of genetics has yet to be rigorously defined. If, therefore, a Darwinian approach to culture is to be of consequence, it must specify how culture is coded and identify the way in which culture is replicated. We have to identify the nature and operation of the units of cultural replication.

The fundamental premise is that cultural phenomena exhibit heritage constraint. They are, at base, replicated according to internal (but not deterministic) code structures or recipes (Fletcher 1977; Dawkins 1986:294–298). These specify the possible and tolerable combinations of signals which a community employs to assemble the features of its behavior. To understand the replication of things or parts of bodies, we must, in both cases, refer back to a complex recipe from which those forms are generated. This requires a description of the signals from which cultural messages are constructed, the nature of those messages, the internal logic of the codes which specify the relationship between the signals of a message, and the way in which the replication works both to retain consistency and to generate variability.

ROLAND FLETCHER • Department of Archaeology, School of Archaeology, Classics, and Ancient History, The University of Sydney, New South Wales 2006, Australia.

Culture consists of several different message systems, each of which uses different kinds of signals. In familiar expression these are referred to as words, actions, and things. These signals are not biological. Nor are they reducible to the biological signals of genetics from which the physical form of the human actors is replicated. A theoretical model of cultural replication has to explain how the different messages constructed from these different forms of signal are related. While that fundamental issue is the primary concern of this paper, the analysis requires a brief discussion of messages composed of things, i.e., material messages, because they are the least understood signal assemblage. The structure of material messages will be illustrated using the specific example of spatial messages made up of the distance signals carried by the buildings in a settlement, e.g., the lengths of walls. The examples are taken from an archaeological context—the two New Kingdom artisans' settlements at Deir el Medina in Egypt (Figure 1) (Bruyére 1939; Cêrny 1965, 1973; Valbelle 1985 for basic information). The durable walling of these settlements allows reliable identification of both the elementary spatial signals and the general characteristics of spatial messages.

Social archaeology has tended to convert material entities into categories of verbal meaning in order to discuss them in "social" terms. The postprocessualists have taken this a step further into contextual analyses of verbal meaning which are unique to each society, as defined by Hodder in 1986. These analyses are modeled on studies of contemporary societies (Robben 1989, Kent 1990), such as the ascription of cosmological content to the form of Batammaliba houses (Blier 1987) and the analysis of Marakwet residence patterns in terms of social contradictions and the role of women (Moore 1986). But these are verbal meanings overlaid on the material. They are not themselves material messages. The material messages carried by buildings are about the actual ordering of space in a community—messages concerning "how far apart," "what degree of visual shielding," or " what distance before a barrier is likely to occur" (Fletcher 1988). Such messages are counterintuitive to a verbally dominated explanatory logic and may seem to have no content. However, much of daily life is a matter of nonconsciously and habitually locating oneself in space relative to objects and people.

Both verbal and material signals alter over time but the latter are not edited out as quickly and completely as the former. The critical outcome is that new socially altering verbal meanings can be attached to old buildings. If there is some kind of coherence in each message system, then this desynchronizing allows that noncorrespondence and dissonance can occur between the different kinds of message. One cannot be regarded simply as a derivative of another. The elaborate, polysemous social meanings which a resident might put on the relative size of the rooms in different houses at any given time cannot be regarded as simply commensurable with the metrical differences in the buildings.

ORGANIZED DISSONANCE

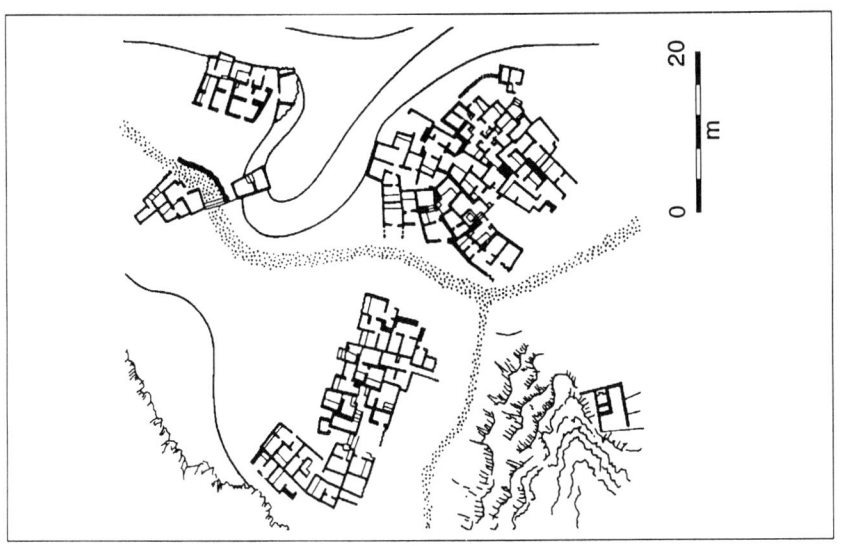

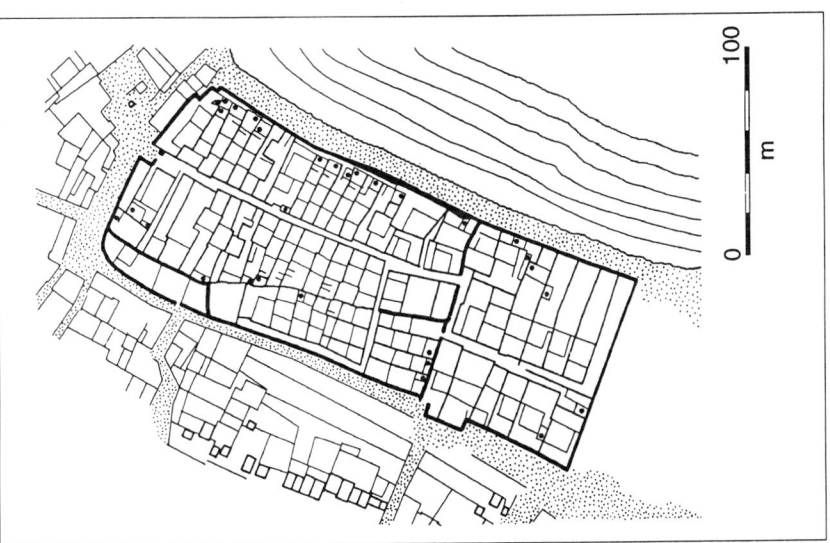

Figure 1. Deir el Medina, Main settlement and Top site.

Verbal meaning is far more dense and versatile than that. Old houses might confer status but smaller rooms could be rather demeaning. The occupants of different parts of a village might even hold two divergent opinions simultaneously and any one person may believe one while declaring the other.

We should distinguish between the differing behavioral algorithms which generate either words, actions, or "things," and between material messages and the varied verbal meanings which the community then allocates to the products of the different message systems. It is a commonplace of social theory that divergences of opinion and social contradictions exist within a community. In addition to the inhomogeneity of verbal meaning I wish to extend the thesis, by arguing that there is also inherent, liable dissonance between the verbal, active, and material components of community life. The implications for a Darwinian model of cultural transmission are considered first, then the broad implications of the model for interpreting the archaeological record are reviewed.

CULTURAL MESSAGES

There are at least three, parallel message systems.[1] Of the three, the first (speech) has been intensively studied in linguistics and semiotics. It is primarily overt and consciously managed. The second (actions) has been clearly recognized since the 1960s in the work on proxemics and kinesics (Hall 1966; Weitz 1984; Poyatos 1988; Campbell 1989). The nature and role of the third (material messages) is rather less understood. Each kind of message, whether verbal, performance, or material is made up of signals—for instance, the words of a sentence, the steps of a dance or the lengths of the walls in a building. There is some ordering to the signals, whether verbal grammars, the sequences and periodicity of movement or the length/width ratio patterns of buildings. These patterns of order vary from community to community. They form distinctive assemblages of cultural features whose transformations through time are characterized by heritage constraint. The replication is imperfect but it is not arbitrary even though the form of the messages changes.

Speech is replicated whenever we speak. Actions are replicated when we gesture or locate ourselves relative to other entities. An assemblage of material signals is replicated when another example is created or an existing feature, such as a building, is altered. Each suite of signals has a different replicative rate and

---

[1] There is at least one other distinct signal suite for color. Sound is also divided into signal systems for speech and music, just as the material includes the shape of small objects and the large-scale patterning of space. By contrast, odor does not appear to have a substantial, systematic role in human cultural signaling.

can endure differentially. They also change at differing rates. To communicate in speech we have to repeatedly and rapidly replicate its signals. Coherent speech has a rapid delivery with short intervals between words. The sounds fade away very quickly. The corollary is that expressions of a spoken language alter relatively rapidly over time because they are repeatedly and rapidly being re-produced. By contrast, even the flimsiest buildings, such as the huts of a !Kung camp or the shade platforms of an Australian Aboriginal settlement, carry a signal for several days about the size of the spaces used by the community. Durable buildings may retain a spatial signal for many years and can carry them for centuries, as in both the Main settlement and the Top site at Deir el Medina. The arrangement of the residence units in the northern part of the main settlement was established in the XVIII dynasty early in the 15th century B.C. The construction of houses in the southern sector began early in the XXth dynasty nearly 300 years later (Bruyere 1939:3–11). Meanwhile, occupation of the older houses had continued without a change in ground level (James 1984:229–233). The extant settlement plan dates largely from the Ramesside period (16th century B.C.). The overall arrangement and shape of the residence units did not change over time though the southern buildings have larger rooms (Figure 2). However, the older spatial message of the northern sector with its smaller rooms was still a component of the inhabitants' perceptual field until the main settlement was abandoned in the late 12th century B.C. (Cêrny 1965:14, 21, 40).

MULTICHANNEL SIGNAL REPLICATION

The various kinds of cultural signals are transmitted at different rates. This is a function of how long the signals of a given type tend to persist and the frequency with which they are replicated (Figure 3). Verbal communication has the fastest transmission rate. Body actions are rather slower, ranging from fast gestures through to the longer durations of positioning, such as sitting in one place. Material entities are replicated over an even wider range of rates, from the relatively fast breakage, loss, and replacement frequencies of some small items, through to the longer persistence and slow replication of durable buildings.

The different channels simultaneously carry information but at different rates and with a time lag between the occasions when their signals are replicated. The lag between the different signals and any lag inherent in replication prevents the attainment of equilibrium (Coveney 1990). We should therefore expect potential dissonance between the different messages, not correspondence, and even some disequilibrium within messages which have slower copying rates. Nonverbal body language decisively belies the notion that verbal meaning is the universal referent for the meaning content of a culture. Inter-

Figure 2. Residence units of northern and southern parts of Main settlement.

ORGANIZED DISSONANCE

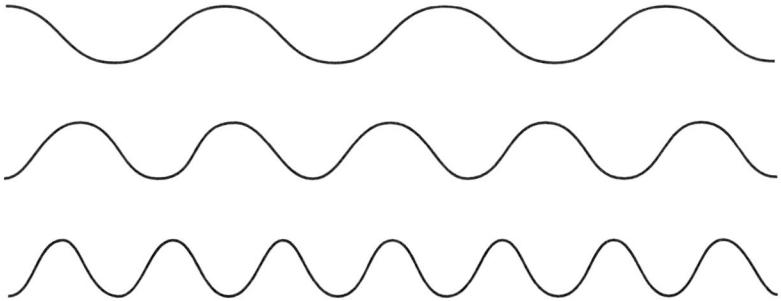

Figure 3. Schematic representation of signal transmission rates for multichannel cultural message systems. Top = verbal, middle=action, lower=material.

personal positioning carries information and meaning without direct reference to verbal meaning and does not require translation into it. Try standing too close to a stranger in a large empty room. We should therefore envisage that all of the cultural message channels carry their own type of meaning content (Fletcher 1989), Interpersonal distances have sensory content . Material spaces have a probability content, concerned with the frequency or predictive likelihood of encountering particular spatial patterns. The meaning content is about the predictability of the sizes of the spaces into which a settlement is divided and its role in the daily life of a community. However, we have tended to presume that verbal meaning is what really matters (Hodder 1986; Spaulding 1988). The erroneous assumption that meaning is exclusively defined by words makes correspondence between the material and the social seem inevitable, since contradiction could only reside within the verbal system, not between it and its context. The notion of parallel communication channels operating at different rates, and not necessarily in continual synchrony, has several interesting implications for the nature of cultural change and the study of the archaeological record. It suggests that the way cultural replication works must be very different from the biological system because the behavior of *Homo sapiens sapiens* operates on multichannel replication of several distinct signal systems, not a single, base code structure as in genetics.[2] We also cannot presume that an "idea-meme" of culture is the adequate definition of a unit of cultural replication (UCR) operationally equivalent to the gene, because some of the cultural units will not remotely resemble "ideas." They will be signals whose meaning content is tacit and takes the form of frequencies and periodicity, not verbal as-

---

[2] Exact equivalence between culture and biology could only be claimed if there is a fundamental, cultural signal system which generates all of the other signal systems from a specific code structure unique to each human group and which is always found in its entirety throughout the replicative system.

sociation. The archaeological record should therefore contain information on aspects of cultural replication that are not explicable in terms of verbal meaning or ideas. This may be of consequence for studies of decision-making and intent of the kind discussed by Mithen (1989) and Shennan (1989). The multichannel model suggests that the slow message systems may define the perceptual milieu of decision-making far anterior to conscious intent, since they would specify the spatial and temporal frame of reference of a community. To understand cognition and rational decision-making would then require an understanding of the material message systems which define how reality is to be perceived, just as Whorf (1953) and Wittgenstein (1953) have argued in different ways for the definitive role of language structure in thought. The conventional, verbal meaning premise of social reconstruction, whether nonprocessual, processual, or postprocessual, must therefore be an inadequate approach to understanding human behavior.

An implication of some general relevance for archaeology is that multichannel transmission at different rates creates potential dissonance between verbal meaning and material context. If "social" is defined in terms of verbal categories of meaning, as it usually is, then we cannot expect stable material–social correlates to exist even within the time span of one contextually unique culture. A critical aspect of the functioning and finite viability of human communities may be that dissonance between the different kinds of messages is a continual possibility. Seeking correspondence between material messages and verbal meaning might actually obscure one of the fundamental processes which both generates change and restricts the potential persistence of human communities.

BASIC PREMISES

We should not make the current complexity of biological theory the standard by which a Darwinian theory of culture is to be judged. Archaeology must perforce proceed through its own theoretical history and should not mimic the contingent history of ideas in another discipline. For instance, we may need to start with propositions which are extremely simple, not because they are true or give especial virtue to their proponents but because they simplify debate. They lead to overt test implications in a manner that is not provided by complex multifactorial propositions. Presumably, we will, in due course, tend towards more elaborate models, as biology has done in its development from the particulate, one-unit genetics of Mendel to the abstruse mathematics, biochemistry, and hierarchical form of the New Synthesis. However, the one addition we do need immediately is hierarchical logic (Grene 1987) in order to deal systematically with the interrelationships between operations and phenomena which display patterning over different spatial and temporal scales (Bailey 1983; Rapoport 1988; Rindos 1989:22; Fletcher 1992; Knapp 1992). In the main, cultures have been regarded either as "organisms" or as sets of traits, the former somehow replicating themselves, and the later being viewed primarily in terms of the adaptive potential of their human possessors (Boyd and Richerson 1985; Leonard and Jones 1987; O'Brien and Holland 1992). I do not propose to use either of these views. Instead, a more appropriate model is provided by viral replication (Cullen 1993). As Dawkins has remarked, "a cultural trait may have evolved in the way it has, simply because it is advantageous to itself" (1976:214). In the virus model, cultural messages, like biological viruses, are copied by exploiting the replication system of another entity. They are copied through the repetition of electrochemical signals in the brain and by repeated motor actions such as speech or activity. Just as viruses are neither alive nor dead, so cultural messages are neither exclusively self-replicating nor a dependent derivative. They exist in both states, within the biological mechanism of the creature and in other forms external to it. Their ready transferability between the humans who replicate them prevents deterministic dependence.

The relevant entity that is subject to selection must be some aggregate of cultural features, whether viewed as a message or an entity, not the biological entities (the humans) who participate in replicating those features. We cannot assume that traits become more or less common simply as a corollary of the biological survival of their human transmitters. A cultural feature could become more common in a human community without affecting either the number of people in it or the reproductive success of the people who carry it. It would be displaying the effects we expect from selection but without altering the reproductive success of its recipients or its transmitters. For instance, biological inheritance is not apparently relevant to the spread of popular music.

THEORETICAL FRAMEWORK

A cultural process equivalent in its operation, structure, and parameters of the new synthesis of the Darwinian model of biological evolution, requires three key characteristics. The fundamental principle is that any signal system evolves by the differential survival of replicating entities (Dawkins 1976:206). Because the resulting forms may either function well or badly in the external world, they cannot derive their characteristics from their eventual function. My concern in this paper is to ask how the signals are produced and organized to be available for selection by external factors to have an effect.

1. A replicating signal system must possess internal coherence; otherwise it could not produce a consistent repetitious pattern through

time (i.e., display heritage constraint) and be independent of determination by the external environment.
2. The process of replication must, however, result in some degree of "error" which generates variability. The replicative code cannot, therefore, be a set of prescriptive and invariant rules.
3. Replication must also be liable to produce more entities than the local environment, whatever it is, can carry. The range of variants will then be subject to selective pressure in a milieu that affects the degree to which cultural entities can persist and differentially replicate.[3]

Two major classes of constraint, the internal and the external, affect an assemblage of cultural signals. The first major class of constraints will be a consequence of internal code consistency which, over the short term, delimits the degree of variation coherent with the existing message structures of the community. This internal selection should operate against signals which do not match the tolerances of the code. *Sorting* is the term used for the genetic equivalent by Vrba and Gould (1986). Some degree of code coherence would be the internal boundary condition to the continued viable replication of a cultural message. Second, there are the constraints of external circumstances which, in the long term, set the boundary conditions within which the products of the replication system succeed or fail. Several levels of external selection exist. Over relatively long time spans, communities are subject to the selective pressures of energy input–output balance by economic and ecological factors. They constitute the boundary conditions which define whether or not a pattern of behavior is viable. For instance, a spatial pattern carried by the community may demand more timber than natural regeneration can supply. If the spatial pattern is highly restrictive, it may commit the community to excessive depletion of its resources for long enough to cause an irreversible decline in supply. Alternatively, if the pattern permits substantial variation, then the community may be able to shift to a less damaging construction policy.

Over shorter time spans, communities are also subject to the boundary conditions set by the pressures of interaction and communication stress (Fletcher 1981a). If behavioral features, such as the arrangement of walls, serve to manage social stress, then the community can continue to function. However, should the community be unable to develop such controls from its behav-

---

[3] My concern in this paper is primarily with what happens prior to selection and adaptation in the external environment, though I briefly discuss the issue. The convoluted difficulties of identifying "adaptiveness" are well illustrated by O'Brien and Holland (1992). The notion of "surplus" signals is not difficult to apply to speech and to nonverbal body language, nor is there much difficulty with applying it to the production of stone tools or pottery. Ironically, given my primary interest, it is much harder to envisage in terms of built space though it applies quite well to the diversity of signals that result from the perceptual disintegration of a community's spatial message.

ioral repertoire, or cannot adopt them from elsewhere, the means to cope with the stress will not be available and the community should be liable to break up. We cannot assume that communities will inevitably devise adequate *in situ* solutions to their problems. The internal logic of their behavior may prevent the creation or adoption of a viable alternative.

THE ELEMENTARY STRUCTURE AND OPERATION OF CULTURAL REPLICATION

From the elements of the operational theory and the basic characteristics of culture, a model can be proposed which incorporates external selective pressures, internal sorting of signals, and the existence of replicators which produce variability in the signal output.

Elementary Code Structures and Message Coherence

Differing signal magnitudes, such as word lengths, the duration of single actions, and the distances which define the positions and sizes of things are the elementary components of the cultural message codes of speech, actions, and material entities. Elementary codes or recipes organize the relationships between discrete signals, to produce cultural messages. Each message has its own discontinuous suite of signals, such as sounds, syllables, and words. Spatial patterns consist of distances which aggregate into discontinuous size frequency distributions and into spatial signatures which are collections of frequency distributions (Figure 4i, ii). Concentration around particular distance values occurs very clearly for some classes of distance. In addition, the same modal values may occur for two different classes of distance. In Munyimba, a Konkomba village in Ghana, this occurs for functionally unrelated entities (Fletcher 1977:80–81). The form of a spatial message is not, however, a simple metrical harmonic—though that is initially a convenient way to envisage it. Concentration and scattering of distance values is present in Deir el Medina suggesting that a single "rule" does not exist for all of the construction events in the settlement. Instead there appear to be probabilistic ranges of options and a tendency for a well-defined pattern to appear only as sample size increases. Pattern forms out of many repeated events but any one event, e.g., the building of a wall, may be unpredictable. At base the occurrence of each distance value is indeterminate. Its precise value cannot be specified beforehand. Instead each is a member of a probability distribution whose form eventually appears but may come about by a different sequence of aggregation even within the various parts of one settlement complex.

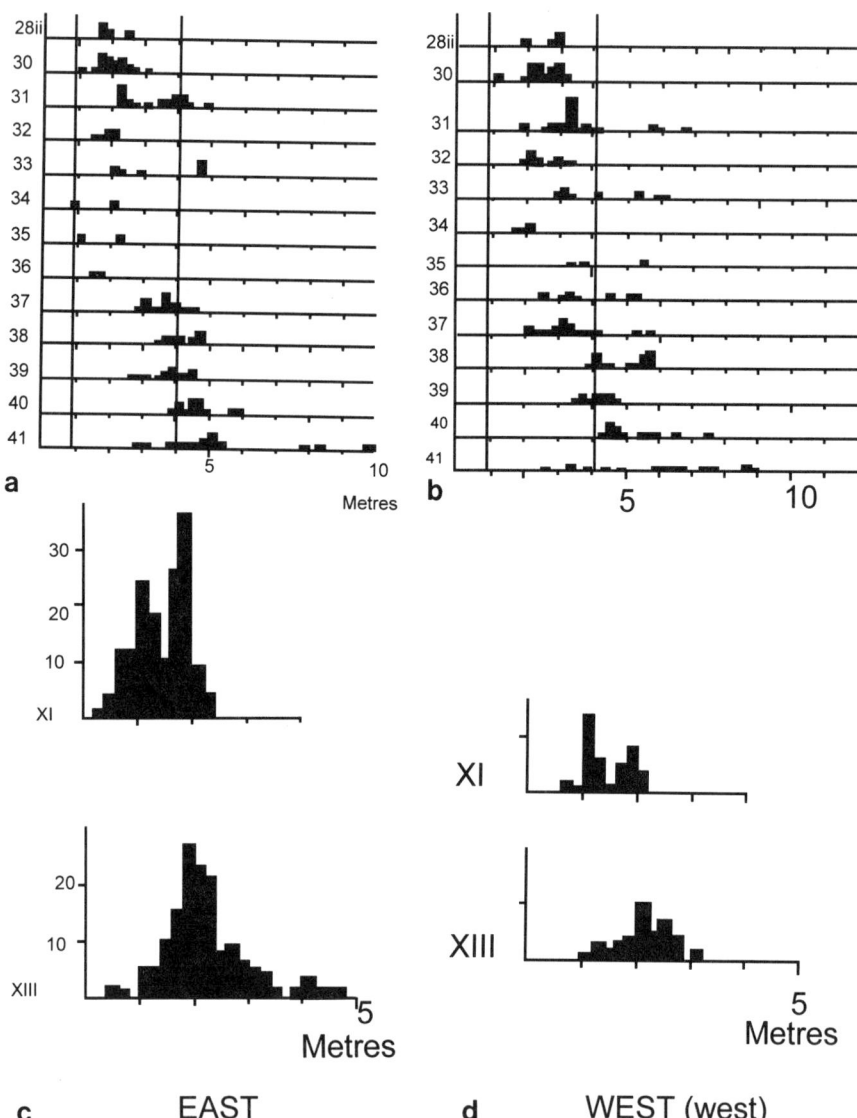

Figure 4. Part of the spatial message for (a) West I and (b) South I in the Main settlement. West I is the northern part of the western extension and South II is the eastern part of the southern extension. Note: numbers refer to different classes of distance, e.g., door widths or broom lengths or room widths. Frequency distributions for room width (XI) and length (XIII) in the (c) East and (d) West sectors of the Top site. Note: the rooms in the Top site are small compared to the diverse range of sizes in the Main settlement.

The essential notion that is required is of a recipe from which the message is generated. As Dawkins pointed out with great acuteness 20 years ago, the analogy between a blueprint and the genetic code is invalid (1976:211). The gene code does not have a one-to-one mapping correspondence with the product. The same genetic material in a different locus in a growing entity will produce a different product. A recipe defines how to make an entity such as a cake, not exactly what it will look like nor that any two versions will be exactly the same. A recipe is a succinct summary of the conditions required to produce an entity given sufficient resources and the appropriate boundary conditions for copying to start. A change in a recipe produces a change in the overall nature of the product, not a change in any one part of the entity. For instance, increasing the recommended temperature may result in a burnt cake. But conversely an alteration to the cake such as cutting a slice out of it cannot be read back directly to the generative code of the recipe. A cultural code should therefore be an algorithm for a product, not a one-to-one description of it.

A recipe, or algorithm, has several important properties. For instance, the same "rules" can result in different products in slightly differing circumstances. It allows what at first sight appear to be markedly different products from one set of signals. At Deir el Medina the same elementary spatial recipe apparently allows two different residential patterns. In simple terms the Top site is the pattern which results from combining the smaller distances of the spatial algorithm, while the order of the Main settlement seems to follow from the prior layout of the streets and a wider repertoire of wall lengths. In essence, code algorithms allow the dense packing of simple referents for a myriad complex possible forms. Not only is clutter kept to minimum, whether in the minds of individuals or the communication system of a community, a variety of products is also possible. Some may, by happenstance, be of potential survival value in unforeseen futures. As Chomsky pointed out in the 1950s, verbal grammars allow the repeated creation of unique sentences which can give new and unexpected meaning to our lives. This capacity cannot be derived simply from experience. Some prior expectation of pattern inherent to the brain and its communicative milieu is required. A recipe is economical of space and detail. The elementary structure of a recipe should derive from the parsimonious storage of information in the brain. We might expect cultural codes to take this form because of the finite storage and analytic capacity of the human brain (Dawkins 1976:211–212; Fletcher 1977:55–60).

A hypothetical model of spatial messages of this sort is provided by a serial ordering of the distances of which residential space is made up. For instance, if settlement space consisted only of distances of either 2, 3, 5, 8 m (i.e., a Fibonnacci series), then that space would be very predictable. More likely in reality is a series in which there is a tolerated range of variability around each median value. Such a series should show up as a succession of modal values each of which is used for several different purposes. We should find that the sizes of entities are produced by a finite repertoire of distances which are used more frequently than others in the behavior of the community as can be seen at Deir el Medina (Figure 4). In this model, spaces are initially generated by a spatial "recipe" and then subject to use which selects against, for instance, functionally inappropriate room sizes. This allows both that communities can create inappropriate room sizes and that they will also tend to tolerate spaces which are only slightly unsuitable. In this perspective the rational use of rooms follows from the existence of an available repertoire, not the sizes of the rooms from a rational appraisal of needs. Room size should therefore be correlated only partially to function and there should be equivalence between the lengths of the walls used in different kinds of rooms, independent of function. Conventional explanations of form in terms of function, actually refer to mechanical function. They are inadequate to explain the nature of settlement space because the communicative function of buildings as carriers of nonverbal messages generated by a spatial code recipe has been neglected.

In practice we will probably find that a spatial recipe is immensely complex.[4] Recognizing spatial messages depends on the class interval which we use. Differing degrees of bunching of signals will be apparent at different grades of observational detail. In addition, human perceptual accuracy decreases with increased distance and a calibrated change in class interval may be necessary to observe comparable degrees of similarity for smaller and larger distances. What this suggests is that the spatial message of a settlement will more properly be described by a diagram resembling an orchestral score of interlinked serial equations. Each will describe a similar structure at a different level of detail, in

---

[4] Actual spatial messages can be expected to be more complex than the simple hypothetical model, as is indicated by the methodology, results, and logical issues discussed in Fletcher (1977, 1981b). Even this crude level of analysis suggested that there are disjunctions within the spatial format of a settlement. Analysis has to proceed from a simple model to identify the gross consequences of spatial patterning which will be readily and obviously observable. We may then have good reason to expend effort on sophisticated methods of periodicity analysis to identify complex formal logic structures in settlement space. The old methodology which I used in Fletcher (1977, 1981b) to try to identify whether or not a serial pattern can be recognized is now inadequate. It was very cumbersome and required prior choices by the researcher about the categories of distance to be recognized. This procedure generates polymodal distributions which another classification might divide by prior categorization, e.g., by defining two classes of room width on the basis of function rather than applying only one inclusive category of room width. A procedure will be needed which can solve the problem of dealing with the small samples which are inherent to spatial patterning in many small settlements. A statistical method for handling the role of small total populations of values is required which does not regard the existing features as an inadequate sample of some hypothetical population. A form of message analysis will be needed which treats the available signals as the total representation of the message available to the communicators. Without this facility, much of the patterning of spatial message analyses will simply fall below sample recognition levels of standard statistical procedure.

effect a fractal structure (Kaye 1989). In due course we should be able to model spatial distance distributions in these terms. In addition we will have to consider whether there is a cultural equivalent of the "Turing patterns" (Cohen and Stewart 1993) which result from the effect of physical boundary conditions on morphogenesis in biological systems.

Sorting in Elementary Codes

The finite capacity of the human brain at any one time should lead to internal selection, by sorting, in favor of parsimonious signals which are most coherent with the prevalent code of the algorithm. We first have to assess how internal selection (sorting) works in cultural signal systems before considering the role of different signal systems with different replicative rates. Claims for internal selection have been turning up in biology for 25 years or more, at least since Lancelot Law White's effort in 1965 to argue for it in genetics. Recently, Edelman has argued for a mechanism of internal selection in neuronal path formation. As Edelman points out, internal selection is not usually the fashionable view (1989:xix), though the code constraint of consistency within the molecular structure of the gene code is now well established. The preference has usually been for some kind of external determination of internal structure, as once was the case in biology with Lamarckism. It persistently appears as an explanation of cultural patterning, as in the adaptationalism of the New Archaeologists. The sensible premise that external reality must somehow determine the internal structure of the code of life forms is fallacious. Likewise the Lamarckian fallacy that the structure of a cultural message, such as the space of settlements, is either determined by or derived from external influences needs to be reappraised.[5]

The key implication of the internal selection, or sorting, model is that we might expect considerable stability in a given set of dimensions, e.g., room width or length, over time. Since the two settlements at Deir el Medina were constructed over several hundred years, they provide a comparative perspective on spatial patterning and development sequences which allows us to see what happens to spatial messages over time. In the Top site, wall abutments al-

---

[5] S. J. Gould is of the opinion (1980:71, 1991:65) that culture is Lamarckian in operation. In my view, Lamarckism is completely excluded, under the Bateson ban (1972:316–333) that copying determined by external demands and needs would lead to such a rapid divergence between the replicated signals that no coherent replicable message could persist. To avoid imputations of a Lamarckian "needs or pressures-driven" theory we have to avoid the terminologies of purposefulness (Dawkins 1976:211). We can get away with lazy expressions in biology which actively ascribe evolution to animals, e.g., "they evolved hooves," because we are unlikely to see various life forms as somehow intending their shape. But we must rigorously avoid prejudging purposiveness of culture until we can see the degree to which it has to be so viewed.

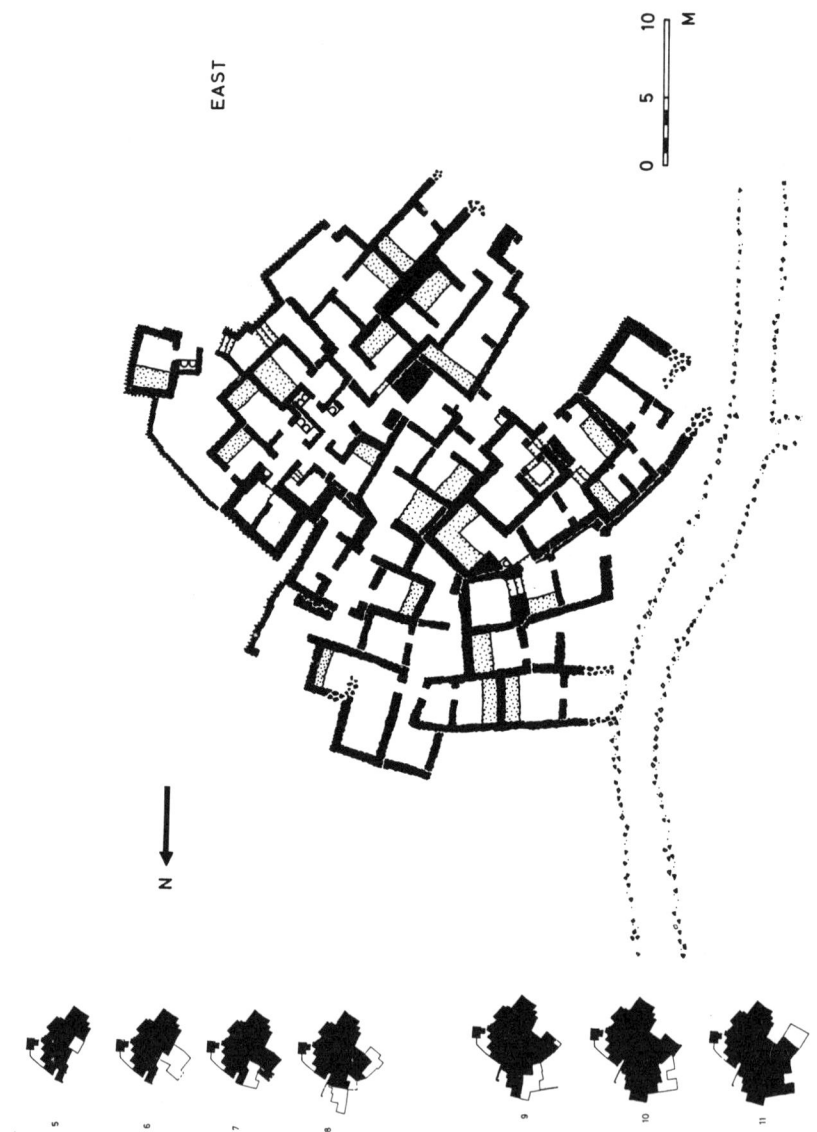

Figure 5. Plan of the East sector in the Top site and its room development sequence.

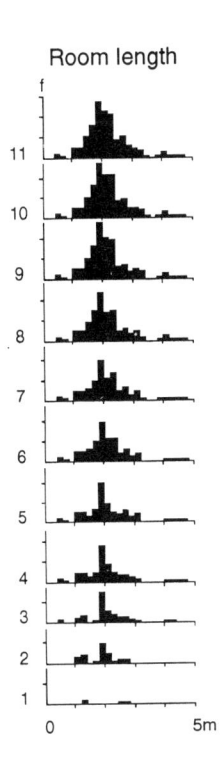 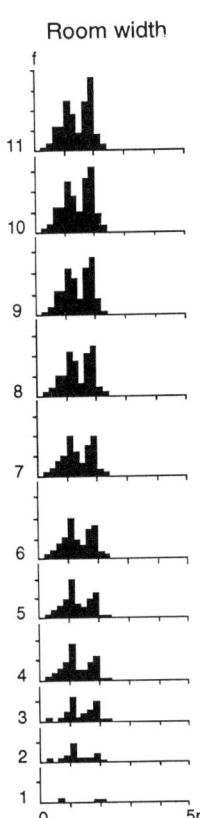

Figure 6. Development sequence for spatial signals of room width and room length in the East sector of the Top site.

low a growth sequence to be identified for each residential sector. The sequence for the East sector provides a clear example (Figure 5). There does appear to be stability in the central tendencies for room width and length, in the form of repetition and consistency over time (Figure 6), suggesting that heritage constraint is operating. However, there is also the potential for new and divergent values to gradually tack onto the limits of the tolerated spread of values. According to the coherence model of a message recipe, any signals that lie within or at the edge of the tolerated range of variability around a central tendency would be acceptable. They could not be excluded by the internal logic of the spatial code because in a large population of distances one marginal addition would have little impact on either the central tendency or the variation of the distribution. The effect can be seen in the Top site development sequences where the adjacent weaker signals were gradually absorbed into the dominant signal. How far this might go is a key issue since it should eventually lead to the obliteration of the internal coherence of the message by merging adjacent signals. The process would initially be slow but as the variability increased it could go faster and faster. Unlike the genetic system, the cultural system apparently cannot stop the internal disintegration of its slow messages. Cultural "degeneration" may actually occur, though ironically not in the moral sphere to which that obsession is usually linked.

THE LOCUS OF THE ELEMENTARY CODE RECIPE

The replication sequences from signal to product differ for each message system. The basic codes and the logic of their combination into cultural signals and messages will be as remote from buildings, dancing and literature as genes and proteins are from hair, hooves, and the panda's thumb. The steps from a recipe to an overt physical, familiar form or a suite of actions are very elaborate. Some equivalence to the succession of copying and shape instructions from DNA to RNA (of various kinds) to proteins and so on, might therefore be expected. Familiar categories such as verbal meaning, e.g., social status and value judgments, are therefore more likely to be the cultural equivalent of eye color, i.e., far along the replicative development sequence, and are not likely to indicate the nature of the fundamental signal components of cultural replication. In verbal communication, permitted sound combinations are arranged into a tolerated spectrum of word lengths, then into utterance or sentence lengths, then up into grammars, and finally to texts of varying degrees of complexity and length. Material space is composed from the probability of occurrence of particular distances. Those probabilities are combined to make enclosed spaces, which in turn aggregate into groups of spaces such as residence units and then into entire settlements for which algorithms of association and route linkage can be devised (Hillier and Hanson 1984). Not only does each signal system have its own distinct replicative rate, it also has its own unique replicative hierarchy.

In whatever way culture is replicated, it involves an interaction between individuals and actual cultural messages. The copying of observed referents external to the human body is an essential component of the process (Fletcher 1977:59–60, 66–67; Sanders 1990; Toren 1990:228–229). Replication does not proceed simply through the internal repetition of disembodied "ideas." The recipe for any given suite of signals simultaneously exists in two forms, one in the transmitted messages and the other in the brains of the observers. For the copying hierarchies to operate, signals have to be produced and then observed by the human sensory system before replication by way of the human brain and the human activity occurs. The role of the human sensory and analytic system as a signal copier is more like that of RNA in the genetic system, as a messenger carrying the signal for new copying, than it is like DNA carrying the primary code. We can therefore escape the tyranny of believing that ideas in the mind are the "template" for the cultural messages. For instance, the total spatial format of a settlement might just as easily be seen as the "template" from which the occupants learn their way of ordering space. Each person does not have to carry all of the algorithm in their head nor does each person have to carry the same algorithm. The notion of the "ideal," whether consensual or socially varied, behind the "Indian" behind the artifact is therefore fallacious. It is founded on a distortion created by seeing human behavior essentially in terms of cognizing—with verbal meanings as the correct and sufficient referent for what goes on in the human mind and as the driver of culture.

Within a community numerous different, partial message structures should reside in an electrochemical form in different people's brains. Each observer's view is incomplete and always will be. There is no monolithic psychological consensus within a community, only an overlapping range of internal variants and an external message. We need not view humans as replicative automata, forever rigidly duplicating their community's standard spatial message. Instead, human beings are creative replicators tacitly and continually relating their internal variants to their partial view of the external aggregate message and creating unique "sentences" or "buildings" as they do so. Some will be consistent with the overall message as others see it and some will not be. Individual idiosyncrasy and overall message consistency are not incompatible. They are just located in different parts of the replicating system.

On this view people create material patterns inadvertently, and then try to do something with them. Our deliberate creative enterprise is to attach verbal meanings to material patterns which are continually and in a sense accidentally coming into being. A more elaborate postprocessual/contextual view is required in which human, cultural replicative activity tacitly creates transformations in deep message structures while simultaneously people play creatively with their verbal meaning structures to obtain purpose and produce social content.

OPERATIONAL CONSEQUENCES

Effects of Noncorrespondence

Because they replicate at different rates, and error is a function of copying, the three codes must generate "error" at different rates and will tend to change at differing rates. *A priori* it is unlikely that each occasion of replication for any one signal class is always accompanied by an exactly consistent shift in the other systems. Therefore, it is not possible for a human community to maintain complete coherence or simple synchrony between the signals or the meaning structures of the three obvious cultural codes. Indeed there should usually be some dissonance between them. The smallest degree of dissonance should occur between phenomena whose replication rates are most similar. Social action and verbal meaning should correspond better with small, rapidly replaced items, such as jewelery, and ought to be expressed in transient decoration, as has been argued by the contextualists (Fletcher 1992:46). Rapidly replaced spatial patterning, as in the camps of mobile hunter-gatherers, should also display links to social organization (Whitelaw 1989). But the "social" should not closely mesh with enduring settlement space and there will not be universal, substantive, uniform connections between them. Trigger recognized an aspect of this more than 25 years ago (1967). Archaeologists practicing social reconstruction have sought a correspondence which should not always be there. To suppose correspondence in order to reconstitute the past is therefore to infringe on one of the fundamental operational characteristics of cultural processes—the potential dissonance between its different message systems.

The social lives of women and their verbal declarations may be seriously at odds with the form of residential space in industrial societies (Ardener 1981; Matrix 1984). High-rise housing has a stressful effect on women with children. We should envisage that societies are affected by a persistent low-level friction between words, actions, and their material context. Not only is society inhomogeneous, its modes of communication are also not in synchrony. What effects this would have in small-scale settlements are hard to define, but residential stress caused by inconvenient route access or overcrowding are possible instances. The industrial urban equivalent is the low traffic speeds of major cities in the late 20th century. Because of the social premises of ownership, the increasing numbers of cars being driven into cities and the severe inertia of 19th century urban space, overall traffic speeds in London will continue to decrease if current policies are maintained (May and Gardner 1990:266–269). In Los Angeles an expanded freeway program will yield reduced average traffic speeds by the turn of the century (Rieff 1992:34). The inertia of the space is being overwhelmed by the activity rates of the community and the declarations of individualism and private ownership (Elton 1992). At its most extreme the dissonance could even lead to catastrophic failure. Ben Elton's "Gridlock"—the ultimate urban traffic jam—is a suitable symbol. But a failure to keep actions in synchrony with material space could occur in a small-scale community whose residential space became so constraining and inconvenient that it made normal interaction highly stressful (Fletcher 1985). We have tended to assume that a community will always be able to change its material context and solve such problems. This presumes that people can correctly diagnose the source of the problem, will overcome the social obstacles to dealing with it, and have the resources to commit to the solution. Our experience in the late 20th century suggests that we should not be quite so confident, whether now or about the past. Cultural collapse deserves the serious attention it has begun recently to receive (Tainter 1988; Yoffee and Cowgill 1988).

What then is the relationship between the different signal systems? Dissonance is consistent with a hierarchical model of adaptation. A hierarchical model is founded on the premise that an entity with survival potential has to be able to make adjustments ranging from the short term and transient, such as altering oxygen intake at higher altitudes, to the long-term irreversible changes of gene frequency in the formation of new species (Bonner 1980:62–63). In culture, words provide the rapid capacity to redefine what people are doing and legitimate a short-term contingent adjustment. In the longer term, changes in residential space slowly alter the fundamental way in which a community manages the spatial dimension of reality. Exact correspondence between the different message systems would throw away the advantages inherent in having several signal systems, each of which allows adjustment to circumstances to occur at a different rate and in a different way. While this can be broadly defined, the specifics of the relationship need much more attention (Fletcher 1988, 1992).

The central implication of multichannel replication is that signal systems will interact through their differing replicative periodicities and occasional unstable fluctuations. The famous "butterfly" effects of dynamical systems theory (Gleick 1988:20–23) may be of great consequence for understanding the range of forms that a cultural system can produce. The combination of recipes and unique initial conditions may help to explain the heritage constraints of culture and our capacity for occasionally creating markedly novel products. The varying periodicities of the different codes may also be able to repeatedly force each other into new variants fast enough to even look like a Lamarckian response to external pressures. I suspect that a sophisticated use of dynamical systems theory will be needed to deal with effects such as period forcing, where the frequency of one signal system acts to stabilize another signal frequency (Gleick 1988:275–300); catastrophic cascades where different periodicities coincide and generate massive unstable change (McRobie and Thompson 1990); and the effects which occur when tiny differences in initial conditions lead to divergent products from the same cultural recipe. Combining the concept of message dissonance with the descriptive tools of nonlinearity might perhaps make the apparently immense complexity of culture amenable to pragmatic analysis.

Variation and "Error"

One profound difference between genetic and cultural replication is that the latter has no capacity to prevent copying divergence from cumulating in the slow, material message systems. This should, in principle, confuse and then wipe out the message structure. What we might therefore find is a gradual increase in variation values as the number of buildings in a settlement increases (Fletcher 1977:86) with a sudden, and terminal, flurry of copying divergence as the various human observers begin to see markedly different spatial patterns in their local portion of the settlement. If coherent spatial messages are needed to keep the framework of active behavior predictable, viable social life should then become rather difficult. We might expect fissioning of the community. Serious attention should be paid to those cultural mechanisms which seek to control replicative variability and their relationship to maximum operable social populations and settlement sizes. The significance for cultural adaptation is that each fissioning group will carry its own fragment of the diversity of spatial signals that the settlement was carrying. As each group begins to build again, it will presumably create a new consistent variant. Not only should this be visible in the archaeological record, it is also a critical mechanism for generating new diverse spatial messages which are then subject to external selective pressures. According to this view, the cloning of material messages is avoided because diversity is cumulated and then splits into numerous variants. Most of those variants are likely to fail in their early stages, e.g., when a weak spatial message is carried by only one or two buildings and a community size is very small and biologically fragile.

The Nature of the Units of Cultural Replications (UCRs)

Dawkins's emphasis on verbal meaning content, e.g., God or celibacy (1976:210), to typify UCRs and this tendency to regard adoption in terms of familiar psychological states such as preference, niceness, and attraction (1976:206, 208), was unfortunate. The label *meme* tends to be linked with defining UCRs as ideas (the idea-meme), and risks linking UCRs to words as the guide to what is in the brain. Such assumptions are still prevalent (Allen 1989:272–275). But this is far too limited a definition of cultural signals. There are other kinds, as Dawkins recognizes in his question about whether a tune or a phrase of a melody is a meme (1976:209). We should shift away from the verbally laden term *meme* toward some other designation, such as UCR, and envisage these units at many different (and also primarily finer and less familiar) scales of magnitude. As Dawkins pointed out for genetics, and by extension for culture, there is no single standard scale of unit which can be viewed as the unit of replication. If a gene is any entity which has just sufficient copying fidelity to serve as a viable unit of natural selection (1976:210), then spatial analysis strongly suggests that the equivalent cultural units will range between single prevalent distances, as carried by a feature such as a wall, up to entities like buildings whose form can be consistently and repeatedly replicated. The basic units of cultural codes will not necessarily be the familiar expressions of social theory. As noted earlier, other kinds of units must exist at levels of meaning remote from the familiar content of verbal meaning.

In addition to the study of verbal meaning, we now need to add the analysis of nonverbal meaning structures, for example, in terms of frequencies and periodicities in the representation of space and time. Bateson remarked, "Clearly there are in the mind no objects or events—no pigs, coconut palms and no mothers" (1972:242). We therefore cannot regard the retention of a cultural feature or its adoption into a design repertoire as sufficiently explicable in terms of conventionally recognized ideas and meanings. The study of decision-making can be further expanded. What we habitually recognize as decision-making is about what people seek to do with the familiar end products of cultural replication. But the way people act depends on perceptual categories. These certainly precede learning, as well as accompanying it (Edelman 1989:7). How a community represents space and time to itself provides the frame of reference for the way in which its individual members make cogent decisions about the adequacy of action and the expenditure of effort.

CONCLUSIONS

Culture is everywhere in a community but is everywhere partial and indeterminate. There are several kinds of cultural message. Each is located in variant forms both within the human mind and in the perceivable signals of daily life. No universal consistent code or algorithm exists within the human replicators, but each kind of cultural message has its own generative recipe which will produce varying products depending on the initial conditions in which replication starts. Noncorrespondence between different message systems is inherent to the behavior of human communities and is a key source of cultural change.

One contribution of this proposed Darwinian approach is that a vast amount of the replicative "error," recombination, and internal selection which contributes to cultural change occurs at levels of operation far simpler than the elaborate overt expressions of culture. There is an immense stock of data in the archaeological record against which to test propositions about these tacit, unstated, nonverbal levels of patterned behavior, especially for the ordering of space. The excitement of this prospect is that there is really much to discover in the past that will be unusual. It lies at our fingertips and will also be utterly unfamiliar and novel—not reducible to the familiar categories of verbal declarations and logic. We will need new research questions, new notations and taxonomy, new kinds of operational models, and new concepts of the boundary conditions of cultural replication.

REFERENCES

Allen, P.M., 1989, Modelling Innovation and Change, in: *What's New? A Closer Look at the Process of Innovation* (S.E. van der Leeuw and R. Torrence, eds.), Unwin Hyman, London, pp. 258–280.
Ardener, S., 1981, *Women and Space: Ground Rules and Social Maps*, Croom Helm, London.
Bailey, G.N., 1983, Concepts of Time in Quaternary Prehistory, *Annual Review of Anthropology* 12:165–192.
Bateson, P.P.G., 1972, *Steps to an Ecology of Mind: Collected Essays in Anthropology, Psychiatry, Evolution and Epistemology*, Paladin, London.
Blier, S.P., 1987, *The Anatomy of Architecture: Ontology and Metaphor in Batammaliba Architectural Expression*, Cambridge University Press, Cambridge.
Bonner, J.T., 1980, *The Evolution of Culture in Animals*, Princeton University Press, Princeton.
Boyd, R., and Richerson, P.J., 1985, *Culture and the Evolutionary Process*, University of Chicago Press, Chicago.
Bruyére, B., 1939, Rapport sur les Fouilles de Deir el Medineh (1933–35), Troisieme Partie. Fouille de l'Institut Française du Caire, Tome XVI. Institut Français d'Archeologie Orientale de Caire, Cairo.
Campbell, J., 1989, *Winston Churchill's Afternoon Nap*, Paladin, London.
Cêrny, J., 1965, Egypt from the Death of Ramesses III to the end of the Twenty-First Dynasty, *Cambridge Ancient History*, 2nd ed., Cambridge University Press, Cambridge.
Cêrny, J., 1973, *A Community of Workmen at Thebes in the Ramesside Period*, Institut Français d'Archeologie Orientale de Caire, Bibliotheque d'étude, Cairo.
Chomsky, N., 1957, *Syntactic Structures*, Mouton, The Hague.
Cohen, J., and Stewart, I., 1993, Let T equal tiger…, *New Scientist* November 6, 1993:40–44.
Coveney, P., 1990, Chaos, Entropy and the Arrow of Time, *New Scientist* September 29, 1990:39–42.
Cullen, B., 1993, The Darwinian Resurgence and the Cultural Virus Critique, *Cambridge Archaeological Journal* 3(2):179–202.
Dawkins, R., 1976, *The Selfish Gene*, Oxford University Press, Oxford.
Dawkins, R., 1986, *The Blind Watchmaker*, Longman Scientific and Norton, New York.
Edelman, G.M., 1989, *Neural Darwinism—The Theory of Neuronal Group Selection*, Oxford University Press, Oxford.
Elton, B., 1992, *Gridlock*, Sphere Books, London.
Fletcher, R.J., 1977, Settlement Studies: Micro and Semi-micro, in: *Spatial Archaeology* (D.L. Clarke, ed.), Cambridge University Press, Cambridge, pp. 47–162.
Fletcher, R.J., 1981a, People and Space: A Case Study on Material Behavior, in: *Pattern of the Past: Studies in Honor of David Clarke*, Cambridge University Press, Cambridge, pp. 97–128.
Fletcher, R.J., 1981b, Space and Community Behavior, in: *Universals of Human Thought: The African Evidence* (B. Lloyd and J. Gay, eds.), Cambridge University Press, Cambridge, pp. 71–110.
Fletcher, R.J., 1985, Intensification and Interaction: A Material Behavior Analysis of Mug House, in: *Prehistoric Intensive Agriculture in the Tropics*, British Archaeological Reports International Series No. 232, British Archaeological Reports, Oxford, pp. 653–682.
Fletcher, R.J., 1988, The Messages of Material Behavior: An Analysis of Non-verbal Meaning, in: *The Meaning of Things: Material Culture and Symbolic Expression* (I. Hodder, ed.), Unwin Hyman, London, pp. 33–40.
Fletcher, R.J., 1989, Social Theory and Archaeology: Diversity, Paradox and Potential, in: *Australian Reviews of Anthropology* (J.R. Rhoads, ed.), *Mankind* 19(1):65–75.

Fletcher, R.J., 1992, Time perspectivism, Annales and the Potential of Archaeology, in: *Archaeology, Annales and Ethnohistory* (A.B. Knapp, ed.), Cambridge University Press, Cambridge, pp. 35–49.
Gleick, J., 1988, *Chaos: Making a New Science*, Heinemann, London.
Gould, S.J., 1980, *The Panda's Thumb: More Reflections in Natural History*, Penguin, London.
Gould, S.J., 1991, The Panda's Thumb of Technology, in: *Bully for Brontosaurus: Reflections on Natural History* (S.J. Gould, ed.), Norton, New York, pp. 59–75.
Grene, M., 1987, Hierarchies in Biology, *American Scientist* 75(5):504–510.
Hall, E.T., 1966, *The Hidden Dimension*, Doubleday, New York.
Hillier, B., and Hanson, J., 1984, *The Social Logic of Space*, Cambridge University Press, Cambridge.
Hodder, I.R., 1986, *Reading the Past*, Cambridge University Press, Cambridge.
James, T.G.H., 1984, *Pharaoh's People: Scenes from Life in Imperial Egypt*, The Bodley Head, London.
Kaye, B.H., 1989, *A Random Walk Through Fractal Dimensions*, Verlagsgesellschaft mbH, Weinheim.
Kent, S. (ed.), 1990, *Domestic Architecture and the Use of Space: An Interdisciplinary Cross-cultural Study*, Cambridge University Press, Cambridge.
Knapp, A.B., 1992, Archaeology and Annales: Time, Space and Change, in: *Archaeology, Annales and Ethnohistory* (A.B. Knapp, ed.), Cambridge University Press, Cambridge, pp. 1–21.
Leonard, R.D., and Jones, G.T., 1987, Elements of an Inclusive Evolutionary Model for Archaeology, *Journal of Anthropological Archaeology* 6:199–219.
May, A.D., and Gardner, K.E., 1990, Transport Policy for London in 2001: The Case for an Integrated Approach, *Transportation* 16:257–277.
Matrix, 1984, *Making Space: Women and the Man-made Environment*, Pluto Press, London.
McRobie, A., and Thompson, M., 1990, Chaos, Catastrophes and Engineering, *New Scientist* June 9, 1990:21–26.
Mithen, S.J., 1989, Evolutionary Theory and Post-processual Archaeology, *Antiquity* 63:483–494.
Moore, H., 1986, *Space, Text and Gender: An Anthropological Study of the Marakwet of Kenya*, Cambridge University Press, Cambridge.
O'Brien, M.J., and Holland, T.D., 1992, The Role of Adaptation in Archaeological Explanation, *American Antiquity* 57(1):36–59.
Poyatos, F., 1988, Introduction. Non-Verbal Communication Studies: Their Development as an Interdisciplinary Field and the Term "Non-Verbal," *Cross-Cultural Perspectives in Nonverbal Communication* (F. Poyatos, ed.). Hogrefe, Toronto, pp. 1–32.
Rapoport, A., 1988, Levels of Meaning in the Built Environment, in: *Cross-Cultural Perspectives in Nonverbal Communication*, (F. Poyatos, ed.), Hogrefe, Toronto, pp. 317–336.
Rieff, D., 1992, *Los Angeles: Capital of the Third World*, Chatto and Windus, London.
Rindos, D., 1989, Diversity, Variation and Selection, in: *Quantifying Diversity in Archaeology* (R.D. Leonard and G.T. Jones, eds.), Cambridge University Press, Cambridge, pp. 13–23.
Robben, A.C., 1989, Habits of the Home: Spatial Hegemony and the Structuration of House and Society in Brazil, *American Anthropologist* 91(3):570–588.
Sanders, D.H., 1990, Behavioral Conventions and Archaeology: Methods for the Analysis of Ancient Architecture, in: *Domestic Architecture and the Use of Space: An Interdisciplinary Cross-cultural Study* (S. Kent, ed.), Cambridge University Press, Cambridge, pp. 43–72.
Shennan, S., 1989, Cultural Transmission and Cultural Change, in: *What's New? A Closer Look at the Process of Innovation* (S.E. van der Leeuw and R. Torrence, eds.), Unwin Hyman, London, pp. 331–346.
Spaulding, A.C., 1988, Archeology and Anthropology, *American Anthropologist* 90:263–271.
Tainter, J.A., 1988, *The Collapse of Complex Societies*, Cambridge University Press, Cambridge.
Toren, C., 1990, *The Social Construction of Hierarchy in Fiji*, Athlone Press, London.
Trigger, B.G., 1967, Settlement Pattern Archaeology, *American Antiquity* 32(2):149–160.

Valbelle, D., 1985, *Les Ouvrier de la Tombe. Deir el Medineh a L'époque Ramesside*, Institut Français d'Archeologie Orientale de Caire, Cairo.
Vrba, E.S., and Gould, S.J., 1986, The Hierarchical Expansion of Sorting and Selection: Sorting and Selection Cannot Be Equated, *Paleobiology* 12(2):217–228.
Weitz, S., 1984, *Nonverbal Communication*, Oxford University Press, Oxford.
Whitelaw, T., 1989, The Social Organization of Space in Hunter-Gatherer Communities: Some Implications for Social Inference in Archaeology, Unpublished Ph.D. thesis, Department of Archaeology, University of Cambridge.
Whorf, B.L., 1953, Linguistic Factors in the Terminology of Hopi Architecture, *International Journal of American Linguistics* 19:141–145.
Wittgenstein, L., 1953, *Philosophical Investigations* (transl. G.E.M. Anscombe), Blackwell, Oxford.
Yoffee, N., and Cowgill, G.L. (eds.), 1988, *The Collapse of States and Civilizations*, University of Arizona Press, Tucson.

Part III

PATHS TO REVISIONISM IN CULTURAL–BEHAVIORAL SELECTION
Individuals and Dual Inheritance

Those attempting to focus on the individual, or at least the individual decision-making process, rather than the group, are conforming to the biological definition of the term *adaptation* that requires reference to individual behavior rather than groups. Following evolutionary biologists and ecologists, the whole notion of group adaptation is rejected unless it is conceived of being no more than the summation of individual adaptations. This move away from cultural ecology to evolutionary ecology for archaeological investigation is of substantial importance. As stated in the Introduction, in this approach the "stable until pushed" premise of processual archaeology is replaced by a view that societies are always in a state of readjustment, experimentation, and change because certain individuals within the society will be attempting to manipulate it to his/her own ends.

Those studies which focus on the interaction between social learning, cultural transmission, and biological evolution find their roots in mathematical approaches. This is especially prevalent in studies of the relationship between style and function that follow. More qualitative approaches are also important, especially dual inheritance theory for its investigation of the interplay between the biological and cultural modes of inheritance.

Some of the basic tenets of sociobiology may also be shown to play a significant role in culture change, again, in regards to the role individual motivations play both in the transmission of information and in culture change. Especially in the realm of decisions that are influenced by kinship, decisions often reflect rules of behavior that are directly conditioned by biological distance.

Chapter 6

Kin Selection and the Origins of Hereditary Social Inequality
A Case Study from the Northern Northwest Coast

HERBERT D.G. MASCHNER AND JOHN Q. PATTON

INTRODUCTION

Anthropologists, and archaeologists in particular, have long used concepts derived from the biological sciences. Terms like *evolution*, *adaptation*, *population pressure*, and *carrying capacity* are commonplace in the archaeological literature. Perhaps no better example of this can be found than in anthropological theories concerning the development of hereditary social inequality where most archaeologists writing on the origin of chiefdoms and states have used one or more of these terms borrowed from biology (e.g., Fried 1967; Binford 1969; Carneiro 1970; A. Johnson and Earle 1987). The recent Darwinian movement within anthropology has brought into question many of the assumptions that underlie such an approach. We will examine this challenge, and argue that a scientific study of hereditary social inequality must be modernized and brought into line with recent advances in evolutionary biology and ecology.

HERBERT D. G. MASCHNER • Department of Anthropology, University of Wisconsin, 1180 Observatory Drive, 5240 Social Science, Madison, Wisconsin 53706.   **JOHN Q. PATTON** • Department of Anthropology, University of California, Santa Barbara, California 93106.

CULTURE AS A UNIT OF ANALYSIS

In the early 1960s, the biologist Wynne-Edwards (1962, see also 1986), building on work done two decades earlier by the demographer Carr-Saunders (1922), argued that the size of animal populations was maintained at levels below the environmental carrying capacity to avoid overexploitation of resources. Population control was the result of individuals imposing reproductive restraints on themselves for the good of the local population. He argued that selection occurs at the level of groups. This concept, known as the Wynne-Edwards hypothesis, has had a major impact on theories of culture. It has been used in explanations as far-flung as describing the rise of state-level societies (Sanders et al. 1967) and the function of ritual dances in Highland New Guinea (Rappaport 1968).

Few contemporary evolutionary biologists believe that group-level preservation mechanisms exist. Perhaps the greatest difficulty in accepting group selection is that it would, by necessity, need to be more powerful than individual selection. Individuals who did not self-impose reproductive limits for the good of the group would be favored by natural selection on an individual level (Williams 1971:11–12). The debate over the Wynne-Edwards hypothesis led biologists to conclude that theories using aggregates (in this case populations) as the unit of analysis, are incompatible within the framework of evolutionary biology (Williams 1971).

Closer to home, Murdock argued that belief in the existence of supraindividual mechanisms of human behavior is part of "anthropology's 'mythology'" (1972:21). He argued that:

> culture, social system, and all comparable supra-individual concepts, such as collective representations, group mind, and social organism, are illusory conceptual abstractions inferred from observations of the very real phenomena of individuals interacting with one another and with their natural environments. The circumstances of their interaction often lead to similarities in the behavior of different individuals which we tend to reify under the name of culture, and they cause individuals to relate themselves to others in repetitive ways which we tend to reify as structures or systems.... [As] reified abstractions, they cannot legitimately be used to explain human behavior. Culture and social aggregates are explainable as derivatives of behavior, but not *vice versa*. All systems of theory which are based on the alleged or inferred characteristics of aggregates are consequently inherently fallacious. They are, in short, mythology, not science, and are to be rejected in their entirety—not revised or modified. (1972:19)

With this strong rejection of theories based on aggregate behavior, Murdock brings into question the validity of the bulk of social science theory. This fundamental error in selecting aggregates as units of analysis can be traced to the founding fathers of the social sciences. Spencer addressed this assumption directly but opted to study the social aggregate, in part because the individual as a unit of analysis had already been taken by the then emerging discipline of psychology. Durkheim and Tylor followed his lead without questioning his decision, or his notion of the "superorganic." Despite a lack of success in formulating any aggregate-based theory with discipline-wide acceptance, most American anthropologists of the 1950s to 1970s never challenged this assumption. Thus, we are left with a body of theory that is based on an assumption that is arguably unscientific, lacks evidence of support from any of the sciences (Homans 1967), and in the case of theories relying solely on notions of population pressure and/or carrying capacity, we are left with a body of theory based on a 30-year-old biological concept long rejected by biologists.

It is now quite clear, as Williams (1966) has demonstrated, that natural selection acts on individuals. There is some argument as to whether or not this can be applied to behavior as well. Alexander (1979) and Boyd and Richerson (1985) have argued that selection can account for variability in cultural systems stating that selection can act on any system of inheritance (Campbell 1975). But the fact that selection is responsible for variability in cultural systems does not remove individuals as the primary source of the variability. As Boyd and Richerson state, "individuals are the primary locus of the evolutionary forces that cause cultural evolution, and in modelling cultural evolution we will focus on observable events in the lives of individuals" (1985:7).

Selection acts on individuals or closely related kinsmen ("kin selection"—Wilson 1975) which is the ultimate unit and mechanism of adaptation (Jochim 1981:17). For cultural behavior, there is agreement that the characteristics of groups must be investigated as the cumulative behaviors of individuals (Vayda and McCay 1975:300; Durham 1976:93; Jochim 1981:17; Maschner 1992) and that individual adaptation can only be considered in terms of decision-making (Jochim 1981:17). Based on Vayda and McCay (1975:302), Jochim argues that we cannot make any assumptions about group homogeneity. Each group is composed of individuals who vary in their abilities, competitiveness, and means (1981).

Jochim believes that groups do have characteristics separate from the individuals which compose them (1981:16). Such characteristics include "diversity, stability, productivity, efficiency, homeostasis, and hierarchical structure... yet nevertheless, an attempt to account for the origin and function of such properties through concepts of group adaptation faces real problems. There is no such thing as group adaptation and all attempts to identify examples of group adaptation have shown that the adaptation is a result of the combined effects of individuals"(1981:16).

Moreover, *ecosystems evolution* has been completely disarticulated. "Natural selection does not act upon entire ecosystems but operates through differential reproductive success of individual organisms within communities" (Pianka 1975:847).

KIN SELECTION AND HEREDITARY SOCIAL INEQUALITY

Explaining the development of hereditary social inequality has long been an anthropological debate which brings out the worst in personalities and often seems to defy common sense. Traditionally, theories of the rise of political complexity have taken a number of forms, nearly all dealing with group-level or population-level behaviors. Social and environmental circumscription (Binford 1969; Carneiro 1970), environmental degradation or improvement (Fladmark 1975), variability in critical resource distribution (Suttles 1968; Hayden *et al.* 1985), or population growth (Cohen 1977; Sanders *et al.* 1979) have been viewed as creating stresses which caused or forced new adaptations to arise (Binford 1983). Trade and exchange (Rathje 1972), intensification of resource procurement (Bender 1981), warfare (Carneiro 1970, 1988; Webster 1975), and the construction of social hierarchies (Ames 1981, 1985; Johnson 1982) are consequences or responses to these stresses created by cultural–environmental imbalances. Because of the difficulties in testing resource stress as a hypothesis, it is usually relegated to an assumption and archaeologists attempt to demonstrate the cause of the stress. Although there has been a strong reaction against population pressure and resource stress models (Cowgill 1975; Hassan 1979; Hayden 1981), the use of population pressure and resource stress as explanatory devices continues today (Binford 1983:197). Overall, we have continuously emphasized the altruistic, the humanistic, the sharing member of society while quietly accepting the fact that everyone is basically working toward their own individual goals. This is seen clearly in Service's explanation for the rise of the state (1975).

In archaeology, this dichotomy is never more prevalent than in Binford's explanation of culture change. He argues that cultures change as a result of the interaction between the environment and the adaptive system. He states that "selection for change occurs when the system is unable to continue previously successful tactics in the face of changed conditions in the environment" (1983:203). Binford claims that population pressure is fundamental to any cultural changes. He argues that culture change is "a function of how people actually solve problems" (1983:221). He believes that the principle of inertia is the most useful construct for looking at change and states that a system will remain basically stable until forces external to the system act on it. Ultimately, it is competition for scarce resources that requires a system-wide shift in adaptation, in our case, the development of political hierarchies.

Yet all of these models have an underlying theme of adaptation and the presumption of adaptive behaviors. This issue has been addressed by Jochim (1981:ix). Jochim states that one of the primary problems in modern ecological anthropology is that it usually addresses the effects of behavior and decisions rather than processes of behavior or decision-making. Thus, it can only deal with optimal behavior, leaving suboptimal, adequate, or maladaptive decisions out of the predictive scenario. He states that an ecological study, contrary to popular belief, must take in more than the physical environment (1981:8). He argues that "an individual's decisions must take into account not only his natural setting but also his family, co-workers and rivals, and the social, economic, and political institutions that define his opportunities and constraints" (1981:8)

Jochim argues that one of the important criticisms of ecological anthropology is that we tend to concentrate on system equilibrium as a foundation for study and give little attention to instability and change (1981:15). I would argue that this is along the same lines as Binford's inertia argument, e.g., nothing changes until pushed (1983). We have no real basis for this argument, it simply makes archaeological interpretation easier. We think that the opposite is probably more true: that systems are always in a state of readjustment, experimentation, and change because the individuals in them, at least some individuals, are always attempting to manipulate the structure in their own, or their kinsmen's, self-interest.

Our thesis is that although populations in adaptation are an attractive unit of analysis in archaeology, they are counterproductive in the sense that evolutionary biologists no longer consider populations a unit of measurement. From a theoretical perspective it is important to recognize that humans act as individuals pursuing goals measured in terms of self-interest. Selection acts on individuals, not populations. Furthermore, if we are to use terms like *adaptation*, we must not distort their meanings, but use them as they are intended to be used. While humans may have behaviors that are adaptive, adaptations, by definition, are a product of natural selection. We must move away from the notion that a population can have an adaptation—it cannot. Although from a theoretical perspective this puts us on firmer ground, it poses difficulties in extending this theory to groups of individuals as any discussion of social ranking must do. Furthermore, such a perspective has little room for altruism.

This is not to say that altruism does not exist, but if we are to use the term *altruism* in relation to theories of adaptation, we must be careful to use it as defined by evolutionary biologists. Altruism exists in two forms, reciprocal altruism (Trivers 1971) and nepotism (Alexander 1974). Reciprocal altruism has been commonly recognized in anthropological writings as "balanced reciprocity." This is not altruism in a pure sense but exchange. The second form of altruism, nepotism, is a behavioral strategy selected for by natural selection to increase inclusive fitness (Hamilton 1964), therefore, not true altruism but rather another form of self-interest. Selection for nepotistic behavior has been dubbed by evolutionary biologists as kin selection (Maynard Smith 1964), and is one of the most important breakthroughs in the study of the evolution of social behavior. Even the strongest critics of Darwinian anthropology have recognized the concept of kin selection in their descriptions of social interactions, albeit unwittingly. Sahlins's analysis of primitive exchange is a case in point. He states:

> The span of social distance between those who exchange conditions the mode of exchange. Kinship distance, as has already been suggested, is especially relevant to the form of reciprocity. Reciprocity is inclined toward the generalized pole by close kinship, toward the negative extreme in proportion to kinship distance . . . close kin tend to share, to enter into generalized exchanges, and distant and non-kin to deal in equivalents or in guile. (1968; quoted in Chagnon and Bugos 1979:213)

Sahlins undoubtedly would frown at the following suggestion, but when he wrote the above passage he was espousing concepts that form the basic tenet of sociobiology.

Kin selection provides a theoretical foundation for a study of inheritable social inequality, properly based at the level of individual status striving. A model based on kin selection would predict that kinship be the basic unit of social organization because it provides a nexus of shared self-interests. Ethnographically, kin-based lineages form a basic political unit. Where more complex political organizations exist, kinship generally dictates membership and often crosscuts and undermines political decisions that run counter to lineage interests.

A kin-selection perspective also leads to certain conclusions about how political units based on lineages would be expected to function, and to conclusions about the origins of hereditary social inequality. First, lineages are, by definition, hereditary units. Membership is a function of descent. Second, because of nepotistic tendencies, individual status would be expected to greatly influence lineage status and *vice versa*. It follows, then, that an important aspect of social status, lineage membership, by definition, is ascribed, not achieved—i.e., inherited. It also follows that hereditary social inequality has it roots in societies well below the social complexity of chiefdoms and states.

Nearly all of the archaeological proselytizing as to the origins of hereditary status differences have been focused on what we call chiefdoms and states, primarily because in this scale of society, status differences are recognizable (Binford 1983; Earle 1989). If someone lives in a stone house on a mound or platform with all of his relatives, and everyone else lives in a thatch hut on level ground, it's a good bet there are status differences. When chiefdoms or states are used as the basis for investigating stratification, hereditary or not, we are explaining social organization in terms of differential material wealth. Traditional approaches have continuously emphasized the role of economic monopolies in the organization and maintenance of hereditary social inequality, but we cannot empirically conclude that economic monopolies gave rise to hereditary social inequality. We need to recognize that there are basic organizational requirements for the creation of monopolies. It would be more proper to argue that the organizational requirements necessary for economic monopolies, hereditary inequality in particular, were in place first in village-based political units—what we traditionally have called tribal societies (Service 1962; Sahlins 1968).

In general, tribal or village-based societies do not usually measure social status by the possession of material wealth. Often status is achieved in the disposal of such material goods, as in the potlatch and exchanges between bigmen. In nonindustrial societies outside of marginal lands, status is more likely influenced by pressures between individuals than between cultures and their environments. The size of one's kin network is difficult to recognize archaeologically, but it is an important criterion for measuring status and prestige. Status based on kin-group size also tends to be inheritable. Ethnographic examples of this are many but data collected by Napoleon Chagnon on the Yanomamo provide an excellent illustration.

The Yanomamo are an egalitarian, sedentary, village-based society, economically based on horticulture and hunting. Villages are palisaded and contain from about 40 to over 250 individuals. These individuals belong to a number of patrilineages, but often over half of the village is a member of a single lineage. Each lineage has a lineage headman. The headman of the largest lineage is, because of the shear number of his political supporters, the village headman. Yanomamo headmen have approximately the same amount of authority as predicted in any introductory anthropological textbook, namely, they work hard to persuade members of their own lineage, and the members of other lineages, to support their aspirations (Chagnon 1975, 1979a,b, 1983, 1988, personal communication).

But under certain conditions, Yanomamo headmen take actual autocratic power. This is done as follows. There is a Yanomamo core area where villages are fairly dense, becoming more scattered with distance from the center. On the periphery of this area, villages are small and fission often. In the center of this area villages are large, often over 200 people. The central villages do not fission as often because small lineages, when on their own, are defenseless against raids from other nearby villages. Under these conditions of social circumscription, a headman can take actual power and express it by taking women from smaller lineages for his own kinsmen and some can even banish individuals from the village, a condition leading to certain death (Chagnon 1975, 1979a,b, personal communication).

These are the conditions necessary for hereditary status differences to develop. An important note is that it is not necessary to have any economic control over production or redistribution, although this may be important in some cases. What is required is a means by which a number of distantly or unrelated corporate groups are forced to live in the same community. It is under these conditions that the headman of the largest lineage *may* take control over social and political decisions.

In applying this model to the archaeological record, the Northwest Coast provides the appropriate prehistoric and ethnographic data.

A NORTHWEST COAST EXAMPLE

Ethnographic Background

On the Northwest Coast of North America are village-based hunters and gatherers who have complex systems of organized warfare (Ferguson 1983, 1984; Kamenskii 1985), institutionalized slavery (Donald 1983; Mitchell and Donald 1986), a hereditary nobility (Oberg 1973; Townsand 1978), part-time craft specialists (de Laguna 1972), and large coastal villages (de Laguna 1983; Townsand 1985). These societies had primarily a marine subsistence base, with a historic emphasis on the massive yearly salmon run (Suttles 1968; Schalk 1977; Langdon 1979).

In southeast Alaska, the region inhabited by the Tlingit, lived people who were semi- to fully sedentary, organized into matrilineages, and where each lineage consisted of 15 to 60 members (usually 20–35) who shared a common dwelling and formed the basic corporate group. At historic contact, lineages were members of clans spread throughout multiple villages, although these were probably single-clan villages prehistorically (Olson 1967; see Acheson 1991 for a similar argument on the Haida). High rank could be bought, earned, won, inherited, or stolen. The maintenance of high rank was based on the splendor of potlatches, political ability, resource ownership, and sheer numbers of lineage members (de Laguna 1983). Lineage houses were internally ranked with nobles at the peak and a middle class made up of kinsmen with decreasing rank as a function of decreasing relatedness. There was a lower class made up of individuals without property or kinsmen who had no rights because they had no support. Below these were slaves who were basically outside the kin-based strata. Lineages and clans were ranked both at the village level and between villages. Lineage houses were ranked within the village and between villages (de Laguna 1983). The headman of the largest lineage in the village was by default the village headman. All economic activities were organized at the lineage level and although some clans held economic properties, their exploitation was generally by individual lineages (de Laguna 1983, 1990).

High rank and status provided, at least historically, a number of benefits, including the control of access to territory and some resources. The individuals who control labor are also the people who are able to mobilize and concentrate people and capital and it is the individuals with greater access to human resources who have a greater say in the decision-making process and will be able to exert their ideas and motivations on others. Since the interests of the elite may be very different from the interests of the population in general, a system-wide measure of maladaptive behavior may be evident for the population in general since the decision-making power is in the hands of only a few individuals (Jochim 1981:105). This is especially true on the northern Northwest Coast where one house might be experiencing a period of bounty while a neighboring, but unrelated house might be experiencing starvation—"gormandizing and starvation coexisted" (Coupland 1988b:240).

Rank was not simply economically advantageous. One of the benefits of high rank was access to others of high rank as spouses. This created political alliances that provided access to other high-status clans of an opposite moiety. Further, the marriage rule encouraged nobles to manipulate their kin relations so that their offspring were not only their biological relations but their lineage members as well—a situation unavailable to other members of society (Rosman and Rubel 1971). Noble males were polygynous and, given the marriage rule, were in a position to increase the size of their lineage over that of lineages of lesser rank. The largest lineage always being the highest ranking lineage.

These societies lived in an environment so rich that Hayden has convincingly argued that there is no possible means, given the aboriginal technology, that they could have ever put pressure on their resource base (1981), and this has been demonstrated archaeologically as well (Maschner 1992). The one primary difference between Northwest Coast societies and the Yanomamo discussed above, is that on the Northwest Coast, resources are differentially distributed in time and space. Under these conditions, peoples tend to concentrate around quality resource locales creating a sort of social and environmental circumscription that has nothing to do with the environment's capacity to support them. Rather, a situation exists where there are numbers of unrelated corporate groups in the same place. This creates interesting opportunities for the headmen of the largest lineages. Since resources are concentrated in certain areas, there are few points on the landscape suitable for village location that would allow the efficient harvesting of those resources. Thus, what are the conditions, both social and environmental, that force people to live in groups larger than their basic kin unit? This is a fundamental problem to this study.

Northwest Coast Archaeological Example

There have been a number of theories put forth to explain the rise of hereditary social inequality on the Northwest Coast, all of which have some role in the process. Certainly Ames's circumscription arguments based on the natural distribution of most resources are of major importance (Ames 1981, 1983). Population increase as demonstrated and modeled by Croes and Hackenberger (1988; but see also Maschner and Ames 1992) is critical and although they argue that populations reached carrying capacity as early as 3000

B.P., and Keeley (1988) argues that they were still at this level historically, there is little supporting evidence. Intensification is also important and Matson has made a case for its significance in Northwest Coast inequalities (1983) as is storage (Testart 1982) and sedentism (Matson 1985). Control of trade and exchange (Bishop 1983, 1987), specialization (Burley 1979) as well as control over key resource locales (Donald and Mitchell 1975; Langdon 1977, 1979; Coupland 1985b, 1988a) have obviously played an important part in the manifestation of inequalities. But contrary to these arguments, there is little evidence that any of these factors were more than conditions or symptoms of social hierarchies, not causes.

We have ethnographically demonstrated that lineage size is the underlying foundation for ranking of the northern Northwest Coast and the only means by which archaeologists on the Northwest Coast can measure lineage size is through house floor area. This is clear in historic villages. In both Tsimshian and Haida cases, the largest house in the village is the house that was occupied by the highest-ranking lineage in the community (McNeary 1976; MacDonald 1983).

Archaeologically this pattern is quite clear. Coupland (1988a:270) found that a historic ratio of large to small houses of 1:3 to 1:5 (ranked to nonranked houses) in a Niska village (McNeary 1976:128) was similar to prehistoric villages on the Skeena River with a ratio of 1:4.4. In the Queen Charlottes, Acheson (1991) also noticed the disparity between large and small houses as an indicator of wealth and size, and found a similar ratio.

The first large houses on the northern Northwest Coast occur in the Middle Phase (Maschner 1992) between 3500 and 1500 years ago. Coupland has made a convincing argument that the Paul Mason site, 160 km up the Skeena River from Prince Rupert Harbor, is clearly an early example of a riverine-based Northwest Coast village, like the inland Tsimshian and other groups during the historic period. This site consists of a large, contemporaneous group of house depressions that date to about 2800 B.P. (Coupland 1988a). Of significance to this argument, the houses are rather small, averaging about 35 m^2, or about the residential size of a few small families. Further, these house floors have little variability in floor area, indicating that the household's were about the same size throughout the village. All of the house floors present on the coasts of southeast Alaska, the Queen Charlotte islands, and on northern British Columbia coasts between 3000 and 1500 years ago are small (Maschner 1992; Ames 1993). There are no recorded villages on the northern Northwest Coast, showing clear evidence of large house size and house size variability dating before about 1800 B.P. (the Late Phase). After 1800 B.P., large house-depression villages begin to appear throughout the region. With the first formation of these large villages we also immediately recognize differential household size based on the floor areas of the houses.

KIN SELECTION AND THE ORIGINS OF HEREDITARY SOCIAL INEQUALITY 99

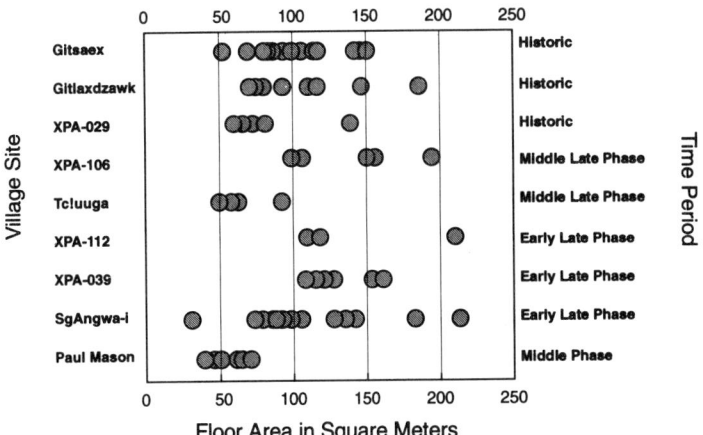

Figure 1. House floor areas for village sites on the northern Northwest Coast spanning the last 3000 years. Middle Phase 3000–1800 B.P., Early Late Phase 1800–1200 B.P., Middle Late Phase 1200–600 B.P., Historic is since A.D. 1770, approximately 200 BP. Data are from Maschner (1992), Acheson (1991), and Coupland (1988a).

For example, in the prehistoric Tebenkof Bay villages mapped between 1988 and 1990 (Maschner 1992), there are distinct size differences between small and large houses. The ratio of large to small houses ranges from 1:2 (XPA-039, -106), to 1:4 and 1:5 (XPA-106 and XPA-029, respectively). When compared with ethnographic villages from the Tsimshian area (northern British Columbia) and the Haida region (Queen Charlotte Islands), the patterning is clear (Figure 1). Moreover, Figure 1 also shows the disparity between the house floors of the Paul Mason Site (3000 B.P.), which was probably egalitarian, and all of the other sites which postdate 1800 B.P., which show clear differences in floor area, and thus household size.

Thus, sometime after 1800 years ago something occurred that inspired a change in the organization of these societies. Houses got much bigger and larger and, for the first time, large villages formed with multiple, and probably unrelated, corporate groups. There is no associated change in subsistence during this transition (Maschner 1992). Maschner has previously argued (1990, 1991, 1992) that this was a period of intense conflict on the north coast, especially after the introduction of the bow and arrow. It appears that large villages of corporate groups formed as a response to conflict and this created a situation where, for the first time, the headmen of the largest lineages were able to put themselves in positions of power and get away with it.

DISCUSSION

In societies that would give rise to complexity, a hypothesis to explain this change must have a component of common sense. It is difficult to imagine a scenario that has a headman controlling an economic resource prior to any other form of social control. The basic social mechanisms that allow individuals to justify economic control in chiefdoms and states were created at the village level. The foundation of inheritable social inequality is in small-scale, kin-based societies. The development of an autocratic power base in these groups is based on the ability to coerce and the ability to coerce is founded in numbers of kinsmen. The ability to organize and lead a group of kinsmen is founded in the concept of the corporate economic group, which is more scientifically valid than the study of cultures or societies (Hayden and Cannon 1982). Corporate groups are a catchall for kin-based reproductive, economic, and political units that are organized in order to facilitate the acquisition of mates or commodities for the good of the kinsmen involved. The formation of corporate groups is probably the fundamental organizational change that leads to hereditary social inequalities.

One method to identify corporate groups archaeologically is with house size or compound size. We do not imply that all corporate groups live under the same roof, but in many cases they do have corporate dwellings. Several authors have remarked that, at least for sedentary villages, there is a clear relationship between house size and social rank (Naroll 1962; Kramer 1979; Coupland 1988a,b). On the Northwest Coast the formation of corporate groups is seen archaeologically by house size. As early as 3000 B.P. on the Skeena River, Coupland has recognized that houses appear to be larger than single-family dwellings, which may indicate the formation of corporate groups (Coupland 1985a, 1988a). By 1800 B.P. houses are in the range of 40 to 50 individuals. Villages show clear size grades of floor areas—the basic measure of rank on the northern Northwest Coast with the largest matrilineages owning the largest dwellings. Since rank was often expressed between lineages in the same village, competition for status was common. The fundamental basis for differential status was numbers of lineage members—the largest lineage was most often the highest-ranking lineage. Thus, headmen were constantly striving to increase their numbers of kinsmen because kinsmen are power.

Conflict has often been forwarded as an explanation of village formation, the development of hereditary social inequality, and the rise of states. In almost all cases this theory is forwarded as warfare solving an adaptive problem. In archaeology this creates a rather entertaining tautology. If we can identify resource stress in the archaeological record, then there was probably conflict, and if we find evidence for warfare, then there assuredly was resource stress. This again takes us to chiefdoms and states. In large-scale societies economic control is the basis for all inequalities. But in small-scale societies, conflict is directly related to striving for status because higher status draws more kinsmen and facilitates access to mates from other powerful lineages. This is the foundation for village-based inequalities because in these societies, all power is based on sheer numbers of kinsmen. Warfare, because of competition for kinsmen and in the defense of status, is probably the ultimate explanation for the development of multi-kin-group sedentary villages. We see the development of sedentary villages as the only basic necessity for the formation of hereditary status differences. Once independent kin groups are forced to reside in the same place, the opportunity exists for the headmen of the largest lineages to put themselves in positions of authority and get away with it. This lineage-based status is a fundamental tenet of kin selection and is the crux of this study.

Yet how will kin selection, translated into lineage size, be seen archaeologically? Status and wealth are often seen in the differential size of houses (Netting 1982:641) and this has been born out repeatedly in the ethnographic record (Cook and Heizer 1968; Kramer 1979; Watson 1979; Hayden and Cannon 1982; MacDonald 1983). Archaeologically, house floor area has been seen as an indicator of wealth and status on the Northwest Coast (Coupland 1985a, 1988a; Maschner 1990, 1991, 1992), as well as in the Southwest (Lightfoot and Feinman 1982), and on the Plateau (Ames 1988).

But this may be more problematic in other areas. Among the Yanomamo the formation of corporate groups is seen in the organization of the villages, the largest patrilineage having the largest section of the village. In some villages, lineage areas are discrete enough to discern areas of occupation. Yet as in most areas of the Amazon, the majority of the artifactual remains are organic. This trait, combined with the rejuvinative characteristics of the tropical forest, makes the archaeological identification of lineage-based status problematic.

CONCLUSION

In all societies there are some individuals who strive for status and high status is culturally defined as to what the society in question considers important (Goldschmidt 1991; Maschner 1992). We argue that hereditary social status will develop everywhere the social and environmental circumstances will allow it. But this behavior need not be adaptive to the society as a whole. It simply needs to not be maladaptive. Maschner has previously argued that ranked societies developed on the Northwest Coast because it was adaptive to be of high rank and yet it was not adaptively expensive to be a member of the middle class (1990, 1992; as has Hayden *et al.* 1985, for the Fraser Plateau). In none of these situations is there any recognizable form of subsistence stress, yet as archaeologists we would be expected to look for deprivation and stress if we are to argue that culture change was occurring. More likely, we should expect that individuals in a culture will be continuously attempting to modify their social landscape in order to better themselves and their close kinsmen. This is a fundamental tenet of evolutionary biology and a product of our evolutionary history (Tooby and Cosmides 1989).

Our job as archaeologists is not to identify individual selection, individual striving, or kin selection in the archaeological record. Our job as scientists is to recognize that they play a critical role in the development of human behaviors and ultimately the development of hereditary social inequality. These are basic, empirically verifiable tenets of human psychology that may be used as assumptions in the study of human interaction (Barkow et al. 1992). It must be assumed that social ranking and political complexity will arise, whenever and wherever it is adaptively possible. It is our job as archaeologists to determine when and where these possibilities occur, what environmental and social variables were significant, and what the prehistoric human actors did about it. Archaeologists must ask themselves why they are studying hereditary social inequality. If it is to better understand the rise of states and chiefdoms, perhaps they should look elsewhere. If our arguments are correct, inheritable social inequality is a basic quality of all political groups and cannot be used as a unique defining characteristic of stratified societies.

Archaeologists who are interested in the origins of hereditary social inequality for its own sake should turn to the study of village-based societies where status differences are founded in the power to control and/or influence kinsmen. An archaeological study of hereditary social inequality should concentrate on the formation of multifamily coalitions or corporate groups, changes in the relative size of lineages, and changes in how lineage-based corporate groups organize themselves in time and space.

REFERENCES

Acheson, S.R., 1991, In the Wake of the *ya'aats' xaatgaay* ['IRON PEOPLE']: A Study of Changing Settlement Strategies Among the Kunghit Haida, Unpublished Ph.D. dissertation, University of Oxford, Oxford.
Alexander, R.D., 1974, The Evolution of Social Behavior, *American Review of Ecological Systematics* 5:325–383.
Alexander, R.D., 1979, *Darwinism and Human Affairs*, University of Washington Press, Seattle.
Ames, K.M., 1981, The Evolution of Social Ranking on the Northwest Coast of North America, *American Antiquity* 46:789–805.
Ames, K.M., 1983, Toward a General Model of the Evolution of Ranking Among Foragers, in: *The Evolution of Maritime Culture on the Northeast and Northwest Coast of America* (R.J. Nash, ed.), Department of Archaeology Publications No. 11, Simon Fraser University, Burnaby, pp. 173–184.

KIN SELECTION AND THE ORIGINS OF HEREDITARY SOCIAL INEQUALITY

Ames, K.M., 1985, Hierarchies, Stress, and Logistical Strategies among Hunter-Gatherers in Northwestern North America, in: *Prehistoric Hunter-Gatherers: The Emergence of Cultural Complexity* (T.D. Price and J.A. Brown, eds.), Academic Press, New York, pp. 155–180.

Ames, K.M., 1988, Storage, Labor, and Sedentism in the Interior Pacific Northwest, Paper presented at the 53rd annual meeting of the Society for American Archaeology, Phoenix.

Ames, K.M., 1993, The Archaeology of the Northern Northwest Coast: The North Coast Prehistory Project Excavations in Prince Rupert Harbor, British Columbia, Report on file at the Archaeological Survey of Canada, Ottawa.

Barkow, J.H., Cosmides, L., and Tooby, J. (eds.), 1992, *The Adapted Mind: Evolutionary Psychology and the Generation of Culture*, Oxford University Press, Oxford.

Bender, B., 1981, Gatherer-hunter Intensification, in: *Economic Archaeology* (A. Sheridan and G. Bailey, eds.), British Archaeological Reports International Series No. 96, Oxford, pp. 149–157.

Binford, L.R., 1969, Post Pleistocene Adaptations, in: *New Perspectives in Archaeology* (S.R. Binford and L.R. Binford, eds.), Aldine Press, Chicago, pp. 313–341.

Binford, L.R., 1983, *In Pursuit of the Past*, Thames and Hudson, London.

Bishop, C.A., 1983, Limiting Access to Limited Goods: The Origins of Stratification in Interior British Columbia, in: *The Development of Political Organization in Native North America* (E. Tooker, ed.), Proceedings of the American Ethnological Society, Washington, DC, pp. 148–164.

Bishop, C.A., 1987, Coast-Interior Exchange: The Origins of Stratification in Northwestern North America, *Arctic Anthropology* 241:72–83.

Boyd, R., and Richerson, P.J., 1985, *Culture and the Evolutionary Process*, University of Chicago Press, Chicago.

Burley, D.V., 1979, Specialization and the Evolution of Complex Society in the Gulf of Georgia Region, *Canadian Journal of Archaeology* 3:131–143.

Campbell, D.T., 1975, On the Conflicts Between Biological and Social Evolution and Between Psychology and Moral Tradition, *American Psychologist* 30:1103–1126.

Carneiro, R.L., 1970, A Theory of the Origin of the State, *Science* 169:733–738.

Carneiro, R.L., 1988, The Circumscription Theory, *American Behavioral Scientist* 31(4):497–511.

Carr-Saunders, 1922, *The Population Problem: A Study in Human Evolution*, Clarendon Press, Oxford.

Chagnon, N.A., 1975, Genealogy, Solidarity, and Relatedness: Limits to Group Size and Patterns of Fissioning in an Expanding Population, *Yearbook of Physical Anthropology* 19:95–110.

Chagnon, N.A., 1979a, Mate Competition, Favoring Close Kin, and Village Fissioning Among the Yanomamo Indians, in: *Evolutionary Biology and Human Social Behavior: An Anthropological Perspective* (N. Chagnon and W. Irons, eds.), Duxbury Press, North Scituate, pp. 86–131.

Chagnon, N.A., 1979b, Is Reproductive Success Equal in Egalitarian Societies, in: *Evolutionary Biology and Human Social Behavior: An Anthropological Perspective* (N.A. Chagnon and W. Irons, eds.), Duxbury Press, North Scituate, pp. 374–401.

Chagnon, N.A., 1983, *Yanomamo: The Fierce People*, Holt, Rinehart & Winston, New York.

Chagnon, N.A., 1988, Life Histories, Blood Revenge, and Warfare in a Tribal Population, *Science* 239:985–991.

Chagnon, N.A., and Bugos, P.E., 1979, Kin Selection and Conflict: An Analysis of a Yanomamo Ax Fight, in: *Evolutionary Biology and Human Social Behavior: An Anthropological Perspective* (N.A. Chagnon and W. Irons, eds.), Duxbury Press, North Scituate, pp. 213–228.

Cohen, M.N., 1977, *The Food Crisis in Prehistory*, Yale University Press, New Haven.

Cook, D., and Heizer, R., 1968, Relationships Among Houses, Settlement Areas, and Population in Aboriginal California, in: *Settlement Archaeology* (K.C. Change, ed.), Stanford University Press, Stanford, pp. 76–116.

Coupland, G., 1985a, Household Variability and Status Differentiation at Kitselas Canyon, *Canadian Journal of Archaeology* 9:39–56.
Coupland, G., 1985b, Restricted Access, Resource Control and the Evolution of Status Inequality Among Hunter-Gatherers, in: *Status, Structure and Stratification: Current Archaeological Reconstructions* (M. Thompson, M.T. Garcia, and F.J. Kense, eds.), Archaeological Association of Calgary, Calgary, pp. 217–226.
Coupland, G., 1988a, *Prehistoric Cultural Change at Kitselas Canyon*, Canadian Museum of Civilization, National Museums of Canada, Ottawa.
Coupland, G., 1988b, Prehistoric Economic and Social Change in the Tsimshian Area, in: *Research in Economic Anthropology, Supplement 3: Prehistoric Economies of the Pacific Northwest Coast* (B.L. Isaac, ed.), JAI Press, Greenwich, pp. 211–244.
Cowgill, G.L., 1975, Population Pressure as a Non-Explanation, in: *Population Studies in Archaeology and Physical Anthropology*, (A. Swedlund, ed.), Memoirs of the Society for American Archaeology 30, Washington, DC., pp. 127–131.
Croes, D.R., and Hackenberger, S., 1988, Hoko River Archaeological Complex: Modeling Prehistoric Northwest Coast Economic Evolution, in: *Research in Economic Anthropology, Supplement 3: Prehistoric Economies of the Pacific Northwest Coast* (B.L. Isaac, ed.), JAI Press, Greenwich, pp. 19–86.
de Laguna, F., 1972, *Under Mount Saint Elias: The History and Culture of the Yakutat Tlingit*, Smithsonian Contributions to Anthropology No. 7, Washington, DC.
de Laguna, F., 1983, Aboriginal Tlingit Sociopolitical Organization, in: The Development of Political Organization in Native North America (E. Tooker, ed.), *Proceedings of the American Ethnological Society, 1979*, Washington, DC, pp. 71–85.
de Laguna, F., 1990, Tlingit, in: *Handbook of North American Indians*, Vol. 7 (W. Suttles, vol. ed.), Smithsonian Press, Washington, DC, pp. 203–228.
Donald, L., 1983, Was Nuu-chah-nulth-aht (Nootka) Society Based on Slave Labor? in: *The Development of Political Organization in Native North America* (E. Tooker, ed.), Proceedings of the American Ethnological Society, Washington, DC, pp. 108–119.
Donald, L., and Mitchell, D., 1975, Some Correlates of Local Group Rank Among the Southern Kwatkiutl, *Ethnology* 14(4):325–346.
Durham, W.H., 1976, The Adaptive Significance of Cultural Behavior, *Human Ecology* 4:89–121.
Earle, T., 1989, The Evolution of Chiefdoms, *Current Anthropology* 30:84–88.
Ferguson, B., 1983, Warfare and Redistributive Exchange on the Northwest Coast, in: *The Development of Political Organization in Native North America* (E. Tooker, ed.), Proceedings of the American Ethnological Society, Washington, DC, pp. 133–147.
Ferguson, B., 1984, A Re-Examination of the Causes of Northwest Coast Warfare, in: *Warfare, Culture, and Environment* (B. Ferguson, ed.), Academic Press, New York, pp. 267–328.
Fladmark, K., 1975, A Paleoecological Model for Northwest Coast Prehistory, National Museum of Man. Mercury Series. Archaeological Survey Papers 43, Ottawa, Canada.
Fried, M., 1967, *The Evolution of Political Society, An Essay in Political Anthropology*, Random House, New York.
Goldschmidt, W.R., 1991, *The Human Career: The Self in the Symbolic World*, Blackwell, Oxford.
Hamilton, W.D., 1964, The Genetical Evolution of Social Behavior, I, II, *Journal of Theoretical Biology* 12:12–45.
Hassan, F., 1979, *Demographic Archaeology*, Academic Press, New York.
Hayden, B., 1981, The Carrying Capacity Dilemma, *American Antiquity* 40(2),Part 2:11–21.
Hayden, B., and Cannon, A., 1982, The Corporate Group as an Archaeological Unit, *Journal of Anthropological Archaeology* 1:132–158.

Hayden, B., Eldridge, M., Eldridge, A., and Cannon, A., 1985, Complex Hunter-Gatherers in Interior British Columbia, in: *Prehistoric Hunter-Gatherers: The Emergence of Cultural Complexity* (T.D. Price and J.A. Brown, eds.), Academic Press, New York, pp. 181–200.

Homans, G.C., 1967, *The Nature of Social Science*, Harcourt, Brace & World, New York.

Jochim, M., 1981, *Strategies for Survival, Cultural Behavior in an Ecological Context*, Academic Press, New York.

Johnson, A.W., and Earle, T., 1987, *The Evolution of Human Societies: From Foraging Group to Agrarian State*, Stanford University Press, Stanford.

Johnson, G., 1982, Organizational Structure and Scalar Stress, in: *Theory and Explanation in Archaeology* (C.A. Renfrew, M.J. Rowlands, and B.A. Segraves, eds.), Academic Press, New York, pp. 389–421.

Kamenskii, A., Fr., 1985, *Tlingit Indians of Alaska*, Translated and Supplemented by Sergei Kan 1986, The Rasmuson Library Historical Translation Series, Vol. II (M.W. Falk, ed.), University of Alaska Press, Fairbanks.

Keeley, L., 1988, Hunter-Gatherer Economic Complexity and Population Pressure: A Cross-Cultural Analysis, *Journal of Anthropological Archaeology* 7:373–411.

Kramer, C., 1979, An Archaeological View of a Contemporary Kurdish Village: Domestic Architecture, Household Size and Wealth, in: *Ethnoarchaeology* (C. Kramer, ed.), Columbia University Press, New York, pp 139–163.

Langdon, S., 1977, *Technology, Ecology, and Economy: Fishing Systems in Southeastern Alaska*, Unpublished Ph.D. dissertation, Department of Anthropology, Stanford University, Palo Alto, Ann Arbor, University Microfilms.

Langdon, S., 1979, Comparative Tlingit and Haida Adaptation to the West Coast of the Prince of Wales Archipelago, *Ethnology* 18(1):101–119.

Lightfoot, K.G., and Feinman, G.M., 1982, Social Differentiation and Leadership Development in Early Pithouse Village in the Mogollon Region of the American Southwest, *American Antiquity* 47:64–86.

MacDonald, G.F., 1983, *Haida Monumental Art*, University of British Columbia Press, Vancouver.

McNeary, S., 1976, Where fire came down from: Social and economic life of the Niska, Unpublished Ph.D. dissertation, Bryn Mawr College, Bryn Mawr.

Maschner, H.D.G., 1990, Resource Distributions, Circumscription, Climatic Change, and Social Inequality on the Northern Northwest Coast, Paper presented at the 6th International Conference on Hunting and Gathering Societies, Fairbanks.

Maschner, H.D.G., 1991, The Emergence of Cultural Complexity on the Northern Northwest Coast, *Antiquity* 65:924–934.

Maschner, H.D.G., 1992, The Origins of Hunter and Gatherer Sedentism and Political Complexity: A Case Study from the Northern Northwest Coast, Unpublished Ph.D. dissertation, University of California, Santa Barbara.

Maschner, H.D.G., and Ames, K., 1992, Prehistoric Population Dynamics on the Northwest Coast of North America, Paper presented at the Annual Meeting of the Northwest Anthropological Association, Burnaby.

Matson, R.G., 1983, Intensification and the Development of Cultural Complexity: The Northwest Versus the Northeast Coast, in: *The Evolution of Maritime Cultures on the Northeast and Northwest Coasts of North America* (R.J. Nash, ed.), Simon Fraser University, Department of Archaeology Publication No. 11, Burnaby, pp. 125–148.

Matson, R.G., 1985, The Relationship Between Sedentism and Status Inequalities Among Hunters and Gatherers, in: *Status, Structure and Stratification: Current Archaeological Reconstructions* (M. Thompson, M.T. Garcia, and F.J. Kense, eds.), Archaeological Association of Calgary, Calgary, pp. 217–226.

Maynard Smith, J., 1964, Group Selection and Kin Selection, *Nature* 201:1145–1147.

Mitchell, D., 1984, Predatory Warfare, Social Status, and the North Pacific Slave Trade, *Ethnology* 23(1):39–48.
Mitchell, D., and Donald, L., 1986, Some Economic Aspects of Tlingit, Haida, and Tsimshian Slavery, in: *Research in Economic Anthropology*, Vol. 7 (B.L. Isaac, ed.), JAI, Greenwich, pp. 19–35.
Murdock, G.P., 1972, Anthropology's Mythology, *Proceedings of the Royal Anthropological Institute of Great Britian and Ireland, 1971*, pp. 17–24.
Naroll, R., 1962, Floor Area and Settlement Population, *American Antiquity* 27:587–589.
Netting, R.M., 1982, Some Home Truths on Household Size and Wealth, *American Behavioral Scientist* 25:641–662.
Oberg, K., 1973, *The Social Economy of the Tlingit Indians*, University of Washington Press, Seattle.
Olson, R.L., 1967, *Social Structure and Social Life of the Tlingit in Alaska*, University of California Anthropological Records No. 26, Berkeley.
Pianka, E.R., 1975, Do Ecosystems Converge? *Science* 188:847–848.
Rappaport, R.A., 1968, *Pigs for the Ancestors*, Yale University Press, New Haven.
Rathje, W.L., 1972, Praise the Gods and Pass the Metates: A Hypothesis of the Development of Lowland Rainforest Civilizations in Mesoamerica, in: *Contemporary Archaeology* (M. Leone, ed.), Southern Illinois University Press, Carbondale and Edwardsville, pp. 365–392.
Rosman, A., and Rubel, P.G., 1971, *Feasting With Mine Enemy*, Columbia University Press, New York.
Sahlins, M.D., 1968, *Tribesmen*, Prentice–Hall, Englewood Cliffs.
Sanders, W.T., Parsons, J.R., and Santley, R.S., 1979, *The Basin of Mexico: Ecological Processes in the Evolution of a Civilization*, Academic Press, New York.
Schalk, R., 1977, The Structure of an Anadramous Fish Resource, in: *For Theory Building in Archaeology* (L.R. Binford, ed.), Academic Press, New York, pp. 207–249.
Service, E., 1962, *Primitive Social Organization*, Random House, New York.
Service, E.R., 1975, *Origins of the State and Civilization: The Process of Cultural Evolution*, Norton, New York.
Suttles, W., 1968, Coping with Abundance: Subsistence on the Northwest Coast, in: *Man the Hunter* (R. Lee and I. DeVore, eds.), Aldine, Chicago, pp. 56–68.
Testart, A., 1982, The Significance of Food Storage Among Hunter-Gatherers: Residence Patterns, Population Densities, and Social Inequalities, *Current Anthropology* 23:523–538.
Tooby, J., and Cosmides, L., 1989, Evolutionary Psychology and the Generation of Culture, Part I: Theoretical Considerations, *Ethology and Sociobiology* 10:29–49.
Townsand, J.B., 1978, Ranked Societies of the Alaskan Pacific Rim, in: *Alaska Native Culture and History* (Y. Kotani and W. Workman, eds.), National Museum of Ethnology, Senri Ethnological Series No.4, Osaka, pp. 123–156.
Townsand, J.B., 1985, The Autonomous Village and the Development of Chiefdoms, in: *Decline and Development: The Evolution of Sociopolitical Complexity* (H.J. Classen et al., eds.), Bergin and Garvey, South Hadley, MA, pp. 141–155.
Trivers, R.L., 1971, The Evolution of Reciprocal Altruism, *Quarterly Review of Biology* 46:35–57.
Vayda, A.P., and McCay, B., 1975, New Directions in Ecology and Ecological Anthropology, *Annual Review of Anthropology* 4:293–306.
Watson, P.J., 1979, *Archaeological Ethnography in Western Iran*, Viking Fund Publications in Anthropology No. 57, University of Arizona Press, Tucson.
Webster, D., 1975, Warfare and the Evolution of the State, A Reconsideration, *American Antiquity* 40:464–470.
Williams, G.C., 1966, *Adaptation and Natural Selection*, Princeton University Press, Princeton.
Williams, G.C. (ed.), 1971, *Group Selection*, Aldine–Atherton, Chicago.

Wilson, D.S., 1975, A Theory of Group Selection, *Proceedings of the National Academy of Science* 72:143–146.
Wynne-Edwards, V.C., 1962, *Animal Dispersion in Relation to Social Behavior*, Oliver & Boyd, Edinburgh.
Wynne-Edwards, V.C., 1986, *Evolution through Group Selection*, Blackwell, Oxford.

Chapter 7

Archaeology, Style, and the Theory of Coevolution

KENNETH M. AMES

INTRODUCTION

Durham's (1991) dual inheritance theory of coevolution provides a coherent basis for applying Darwinian theory to socio-culture change. His theory rests on three hypotheses: (1) Decision making by individuals is the primary but not only cause of cultural evolution. Humans actively choose among alternative courses of action, when they are able; selection is based on evaluation of the consequences of different alternatives. Since knowledge is imperfect, the outcomes will be imperfect. Choices in cultural selection are made on the basis of learned cultural values. Cultural values are socially transmitted between individuals and generations. (2) The relationships between culture and genes are mediated through five modes. These modes constitute coevolution. (3) In general, "cultural variants which improve the reproductive fitness of their selectors will spread through a population by choice or imposition at the expense of alternative variants."

In Durham's theory, cultural selection is the primary force which sorts among cultural variants within a human group. Variants are evaluated against what he terms *primary* and *secondary* values. Primary values are essentially biological—they are the result of natural selection operating on human genotypes. Secondary values are cultural, are learned, and are socially transmitted.

KENNETH M. AMES • Department of Anthropology, Portland State University, Portland, Oregon 97207.

They too change as a result of cultural evolution. Durham's theory integrates biological and cultural evolution without biological determinism.

Coevolutionary theory provides a richer framework for a Darwinian approach to central archaeological issues such as style, than do other, related approaches such as cultural selectionist theories, and other, more narrow applications of natural selection (e.g., Dunnell 1978). This paper provides a sketch of dual inheritance theory, presents an archaeological approach to style based on dual inheritance theory and develops a preliminary application of the approach to the evolution of Northwest Coast art.

DUAL INHERITANCE

The works of Boyd and Richerson (1985) and Durham (1991) are the bases for the following discussion. The reader is also referred to Bettinger's (1990) excellent recent review and exegesis of Boyd and Richerson. There are differences between Boyd and Richerson, and Durham, and between these and similar theories (e.g., Rindos 1984). However, the intent of this paper is to lay out their fundamental similarities, and to explore some implications.

Dual inheritance theories postulate that, in Durham's words, "...genes and culture constitute two distinct but interacting systems of information inheritance within human populations" (Durham 1991:419–420). Both Durham and Boyd and Richerson go to great lengths to establish that genes and cultures are separate systems which are not completely analogous to each other. Following Durham again, the differences between cultural and genetic inheritance include:

> ...first, genes and culture each contain information within codes that have very different biophysical properties (DNA versus memes [see below]); second, the information is stored and processed in different, highly specialized structures (cell nuclei versus the brain); third, it is transmitted through space and time by very different mechanisms (sexual versus *social* intercourse [emphasis Durham's]); and fourth, the information in either system may undergo lasting, transmissible change without there being a corresponding change in the other (Durham 1991:420 [brackets Ames's comments])

In dual inheritance theories, culture is defined narrowly. Boyd and Richerson define culture as:

> ...information capable of affecting individuals' phenotypes which they acquire from other conspecifics by teaching or imitation. (1985:33)

According to Durham, culture has these properties:

1. Conceptual reality. Culture is "shared ideological phenomena (values, ideas, beliefs, and the like) in the minds of human beings. It re-

fers to a body or a 'pool' of information that is both public (socially shared) and prescriptive (in the sense of actually or potentially guiding behavior)" (Durham 1991:3).

2. Social transmission. To qualify as cultural, in Durham's view, information must be learned from others, not genetically inherited or gained through individual experience. In the same vein, Boyd and Richerson also restrict their concept of cultural transmission to situations in which behavior is acquired "directly from conspecifics by initiating their behavior or because conspecifics teach...by reinforcing appropriate behavior" (1985:35).

3. Symbolic encoding. To Durham, symboling is important because its arbitrariness "enhances the information density of social transmission" (1991:6); cultural information so encoded creates realities and organizes relations. Boyd and Richerson do not believe symboling is necessary to their theory.

4. Systemic organization. Culture is treated as a " 'system of knowledge' [Keesing 1974:89] within a population" (Durham 1991:7) that possesses both a hierarchical structure and coherence "(component beliefs are often linked together and embedded within the whole)" (1991:6). It does not follow from this that the components of a culture will necessarily appear coherent to observers from another culture (e.g., Gellner 1988).

5. Social history. The information that constitutes a culture does not emerge fully formed, like Venus from the brow of Zeus. Rather, the information has been handed down, and is composed of "the surviving variants of all the conceptual phenomena ever introduced and socially transmitted" (Durham 1991:8).

In sum, Durham defines cultures as "systems of symbolically encoded conceptual phenomena that are socially and historically transmitted within and between populations" (1991:9).

Behavior is not part of culture in dual inheritance theory; it is rather an aspect of an individual's phenotype, which, among humans, is the result of the interplay among genotype, culture, and the environment. Dual inheritance theory rejects strictly sociobiological explanations of human behavior (Boyd and Richerson 1985; Durham 1991); in other words, behavior is not the direct consequence of genotype. Humans have a genetically based capacity for culture, and cultural variations among human groups are the result of selection operating on cultural inheritance, not on the genotype (Rindos 1986). On the other hand, behavior is not solely the direct consequence of culture either, but the result of many things, of which culture is one.

From the standpoint of an archaeologist, this means that material culture is not culture, but rather the phenotypic expression of the interplay among cultural inheritance, genotype, and environment. This of course is directly at odds with the definition of culture which forms the basis of processual archaeology (Binford 1962), and may seem at first to be a return to an unadulterated idealism. This is not the case. I will explore this issue in terms of style in more detail below. At this point, it is sufficient to say that symbols in material culture—the letters on this page, for example—are parts of a phenotype, in this case mine. Since they are phenotypic, such symbols cannot be seen as the culture code made manifest, a special window into the information systems that constitutes culture. To harken back to an ancient debate in archaeology, artifact form represents behavior, not a mental template.

Durham identifies two basic categories of cultural evolutionary forces: nonconveyance and conveyance forces. Nonconveyance forces include innovation, synthesis, migration, diffusion, and cultural drift. (Recent discussions of style by archaeologists attempting to build a cultural evolutionary theory [e.g., Dunnell 1978] have emphasized the role of cultural drift in stylistic change. That issue will be the focus of the next section.) Nonconveyance forces are the sources of variation within cultural systems. Conveyance forces produce "differential social transmission of allomemes within a reference group" (Durham 1991:422), and include transmission forces, natural selection, and cultural selection.

Before discussing transmission forces, it is important to discuss what is transmitted and who does the transmitting. Durham defines the *meme* as "the variable unit of transmission in cultural transmission. I [suggest] that meme should refer to any kind, amount, and configuration of information in culture that shows both variation and coherent transmission" (1991:422). The "behaviorally expressed variants of any given meme are its 'allomeme.'" Durham argues that memes are imperfectly analogous to genes in biological evolution. (Boyd and Richerson eschew any definition of a unit of transmission. It is beyond the scope of this paper to review the pros and cons of this issue.) Durham believes that the notion of the meme will be among the most controversial aspects of his theory and I suspect the meme concept may be anathema to some. However, the concept is an exceedingly powerful one because it focuses attention on the question of what is being transmitted and on what cultural selection is operating. I will return to this question below.

If memes are the unit of transmission in Durham's theory, then the reference group is the unit of evolution. (In biological evolution, individuals are subject to selection, but they cannot evolve; populations evolve across generations of individuals. Hence, reference groups evolve culturally across generations of individuals.) A reference group is composed of culture carriers with the same range of allomemes, the same range of ecological consequences for those allomemes, and the same value system (Durham 1991:427). The concept of the reference group also allows us to attack the question of what the locus of the culture change is, or as Dunnell has recently asked, "What is it that actually evolves? (Dunnell 1992)"

In coevolution, the reference group is the lowest unit of information above the organism, and therefore, at that basic level, it is reference groups that evolve. However, reference groups are nested within larger-scale "packages" of cultural information, such as ethnic groups, social classes, dialect tribes, and so on.

Two factors make things even more complex. First, culture is self-selecting. As an individual, one can modify, alter, or even abandon values and memes, while one cannot (without modern medical interference) modify one's own genotype. Further, secondary values evolve as they are evaluated against primary values, and others against secondary values. Third, an individual may be a member of more than one reference group at a time. In the United States an individual may make contradictory choices based on gender, ethnicity, race, and socioeconomic class. These give the units of cultural evolution a fluidity and slipperiness which is lacking in biological evolution, though, as Dunnell points out, the issue of the units of selection (i.e., are species "real") has not yet been settled in biology.

However, for coevolution to work, at least one informational level above the individual is required. In the theory of coevolution, culture is transmitted through social learning. The organism therefore must be part of some kind of society, however minimally that society is defined. In Dunnell's view, the major force of cultural evolution has been natural selection acting on individuals, until the development of complex societies within the last 6000 years, when some form of intersocietal selection developed (Dunnell 1980, 1992). From the stand-point of dual inheritance theory, Dunnell is mistaken; social groups and organization have always been important in cultural evolution.

The capacity for culture evolved naturally sometime during the late Pliocene and early Pleistocene, i.e., it evolved under natural selection. If culture is indeed socially transmitted, then it evolved based on preexisting social behaviors. Social behaviors were co-opted by natural selection for this novel form of information transfer; culture must have evolved through natural selection for social learning and social learners. It follows from this that some social behaviors may be "fixed": if they do not occur, learning cannot occur; if learning does not occur, culture is not transmitted. The processes by which language is acquired may be some support for this. From the standpoint of coevolution, then, higher-level social entities must exist for transmission to occur.

The minimal unit within which cultural selection occurs is the reference group; but ultimately it is still individuals doing the selecting. The reference group is itself nested, or embedded, in larger-scale social units, and reference groups can have fluid memberships. This carries the implication that the effects of cultural selection for memes can be at a much larger scale than just that of a particular reference group. This quality may provide us a linkage between *habitus* and large-scale developments.

With this background, it is possible to discuss conveyance forces. Durham divides these into transmission forces: natural selection and cultural selection. Transmission forces arise from the effects of how allomemes are transmitted—socially learned—and the social context in which they are learned. Boyd and Richerson (1985) stress the effects of transmission forces. Of particular interest is their concept of indirect bias transmission. In indirect bias transmission, an individual will adopt a set or complex of traits from a role model on the basis of a single *indicator* trait, or a few such traits, possessed by the role model (Boyd and Richerson 1985). One need only attend an archaeology convention—where academic success is the indicator trait, and the male side of the throng is bearded and be-tweeded—to see biased transmission in operation.

Natural selection is the "natural selection of cultural variation." Durham identifies two modes of natural selection: intrasocietal and intersocietal. In the former, an allomeme aids its bearer to have more offspring than the other allomemes and therefore is learned by more individuals than are other variants. Intersocietal selection "stems from the differential expansion and extinction of whole societies....This force occupies a place in the cultural scheme equivalent to the one species selection occupies in genetic theory" (Durham 1991:431).

Cultural selection is Durham's primary force of cultural evolution. In cultural selection, people select among available allomemes according to the allomemes' consequences. These choices may be conscious or unconscious. Cultural selection is guided by the values held by individuals in a reference group. These values are not necessarily, or often, directly related to biological reproduction. In other words, people do not make most of their day-to-day decisions on the basis of long-term reproductive success. Nor are people necessarily aware of the long term ecological consequences of their decisions (Braun 1991). However, it is a key assumption of Durham's theory that cultural choices will, over the long term, promote reproductive success. There are two categories of the values: primary and secondary. With primary values, evaluation is made directly by the nervous system; with secondary values, evaluation is made against values which are themselves socially transmitted. Secondary values are "*cultural* standards derived from primary values through experience, history and rational thought" (Durham 1991:432). Secondary values themselves evolve.

Food preferences are obvious examples of primary and secondary values. Humans are notorious omnivores, with particular cuisines shaped by "experience, history, and rational thought," but the evolution of these cuisines are shaped by the limits in what we can tolerate in taste, what we can digest, keep down, or what will or will not kill us immediately—primary values.

There are two modes of secondary value selection, or perhaps a continuum between free choice and imposition. In the former, individuals make their own evaluations, while in the latter, the choices of some members of a reference group are imposed on the other members, or the choices of one reference group are imposed on other reference groups through the exercise of power. Most societies fall somewhere in the middle of this continuum. The application of the concept of power is one of Durham's major emendations to dual inheritance theory, and will be discussed below.

CULTURAL SELECTION, CHOICE, AND IMPOSITION

Cultural selection, in Durham's theory, is not a simple analogy to natural selection in biological evolution. In biological evolution, natural selection is the result of differential reproductive success. The relative frequencies of genes and their alleles in a population change in frequency through time according to the relative reproductive success of the bearers of those genes.

Natural selection sorts genotypes according to the differential reproductive success of the phenotypes bearing the genotypes. Different genes and alleles potentially will contribute differentially to that success. This potential is measured by the concept of *genetic fitness*, which is the genotype's *potential* contribution of genes to future generations (Boyd and Richerson 1985; Durham 1991; O'Brien and Holland 1992). As Eldredge (1986) points out, "natural selection is differential reproductive success as determined by economic [read ecological] success—among a collection of (conspecific) organisms" (1986:358). Eldredge views ecology as the economics of nature. In any case, natural (or what Durham calls genetic) selection sorts phenotypes by their ecological consequences.

Durham stresses that cultural selection is "the differential social transmission of cultural variants through human decision making" (1991:198), as people evaluate the consequences of the cultural variants. Durham calls this "selection *according to* consequences." People evaluate the outcomes of the cultural traits against secondary values. These consequences can be ecological, but they can, and more likely will, be social. In dual inheritance theory, culture change occurs as a result of the *differential transmission* of cultural variants.

The *cultural fitness* of a meme or allomeme is "its overall suitability for replication and use within the cultural system of a given subpopulation" (Durham 1991:194). Cultural fitness may have little to do with biological fitness; cultural selection is by evaluation against secondary values, not directly against biological reproductive success. In other words, natural selection can act directly on cultural traits, but not necessarily. However, since the capacity for culture is an evolved human biological trait, it is unlikely that, over the long run, cultural traits will have low genetic fitness values (produce behaviors which lead to low reproductive success). However, cultural selection can contradict natural selection in the short run.

To measure the effect of cultural selection on biological reproduction, Durham uses inclusive fitness (Hamilton 1964). As generally understood, inclusive fitness is a genotype's expected contribution to the next generation both through its own reproduction and its effects on the reproduction (genetic fitness) of close kin. In Durham's version of dual inheritance theory, inclusive fitness is "the average of the individual inclusive fitness of all members of a subpopulation who act on the basis of that allomeme as compared to others...[it] is...a measure of an allomeme's effect upon the reproductive fitness of its carriers" (1991:196–197). It is a key postulate of Durham's theory that cultural selection will favor, over the long run, allomemes that promote the inclusive fitness (biological reproduction) of those allomemes.

The question at the heart of applying the concept of cultural selection is which reference group or subreference group is transmitting the particular allomeme. A particular set of memes may promote the inclusive fitness of a particular reference group within a society, and depress or lower inclusive fitness values for other reference groups in that same society, or the allomemes may be neutral to those other reference groups. Transmission of cultural values promoting high levels of energy consumption in industrialized societies like the United States obviously promotes the inclusive fitness of those societies, but may depress those values in other societies. Actually, the situation may be more complex—the existence of industrialized societies may actually promote the reproductive fitness of so-called third world societies (population levels in those societies are increasing rapidly), but in appalling living conditions. Biological reproductive success is not a measure of quality of life.

A reference group's inclusive fitness could be lowered simply by the presence, or existence, of another reference group with different memes, and the simple existence of this other reference group acts as a constraint on the reproductive success of the first. Or, the inclusive fitness of the first is lowered because the allomemes of the second are imposed on the first. The latter reference group exercises some form of power over the first group, and this power negatively affects the reproductive success of the former. Or, conversely, imposition could raise the inclusive fitness of both reference groups. Imposed cultural selection represents the application of the concept of power to dual inheritance theory.

Power can be classified in several ways. Durham uses Lukes's (1974) classification: power may be coercive (choice is determined by threat), force (choice is eliminated), manipulation (values of subordinate reference groups are controlled), and authority (choices are made because commands fit values of subordinate groups). Yoffee (1991) has classified power on the basis of what is controlled and where in a social system power is applied: social, political, or economic.

Wolf (1990:586) writes of four modes of power. The first "is power as an attribute of a person"; the second is the capacity of an individual to impose their will on another; the third is tactical or organizational power. In this mode, the actor or actors have the ability to circumscribe or control the acts of others in determinate settings. The fourth mode, structural power, is power that structures the political economy; it is power "to deploy and allocate social labor...[s]tructural power shapes the social field of action as to render some kinds of behavior possible, while making others less possible, or impossible" (1990:587).

In both tactical and structural power, some individuals or reference group controls the direction of cultural selection, and the form and direction of transmission. They may control the cultural selection of secondary values, the evaluation of memes against those values, and so forth. They may control the creation and direction of cultural variability, the creation or acquisition of new memes and allomemes. Modern advertising can be seen as the control of culture change by manipulation of social transmission through the production and manipulation of secondary values.

The evolution of the Roman Republic into the Roman Empire may be a classic case of imposition, with important implications for our understanding of large-scale process. According to Millett (1990), the empire resulted from the expansion of the city of Rome from a local center to an imperial city. This expansion was fueled by the internal competition within the oligarchy of the late Republic, in which the success of oligarchs depended on military success, ownership of great tracts of land worked by slaves, the capacity to put on great public displays, and the maintenance of patron–client relationships.

If this interpretation is accepted, then it suggests the Roman Empire was the result of the capacity of a single reference group to impose its culture and interests upon many millions of other human beings rather than the operation of universal social or cultural processes. My point here is to suggest the importance of the concept of power and of imposition for the construction of a coevolutionary theory of human history.

The central question raised by the concept of cultural selection by imposition is who is doing the selection. Who benefits and why? Coevolutionary theory does not assume that culture is a well-integrated whole, functioning harmoniously for the immediate benefit of all. Rather, it wonders who is benefiting at any given moment. It also assumes, however, that, over the long term, the memes being reproduced must benefit the society as a whole, or the memes will, at some point, be selected against. Thus, in coevolution, the expansion of the Roman Empire must, in some way, be demonstrated to have improved the inclusive fitness values of the majority of people who were part of it. This does not mean they were happy, lived long and joyful lives, or engaged in meaningful philosophical discourse. It merely means that for a time, being part of the Roman Empire provided immediate culturally defined benefits and long-term biological benefits,

Millett sees "Romanization" as a key process in the success of the Empire. Roman culture became a "cosmopolitan fusion of influences....[w]e must see 'Romanization' as a process of dialectical change, rather than the influence of one 'pure' culture upon others" (Millett 1990:1). As the Roman oligarchy achieved first tactical and then structural power over larger and larger areas, and more and more groups, local elites probably benefited by achieving local tactical power by taking on elements of Roman culture. Why they benefited is the key empirical question to be asked.

STYLE

Introduction

Artifact form and the relationship between style and function are central issues in archaeology. All attempts at theory building in the discipline must grapple with style. For this reason style already has a central place in recent efforts at developing evolutionary theory in archaeology by Dunnell (1978) and O'Brien and Holland (1990, 1992). However, I think their efforts will produce a terminological and conceptual dead end, and that an approach based on dual inheritance theory will be more productive. I will illustrate my points with a limited discussion of the evolution of Northwest Coast art

Style

In an effort to align archaeological definitions of style and function with evolutionary theory, Dunnell argued that function, in archaeology, is "manifest as those *forms that directly affect the Darwinian fitness of the populations in which they occur*" (1978:199). He defined style as "those *forms that do not have detectable selective values*" (1978:199). Stylistic traits are neutral, neither contributing to nor detracting from adaptedness. Functional traits, according to Dunnell, are under positive selection[1] because they contribute to the reproductive success of the people who possess the trait—whose artifacts perform that function, as I understand him. O'Brien and Holland (1992) seek to clarify and expand his definition to include situations where the possession of a trait contributes to adaptedness, but the trait is not under positive selec-

---

[1] Positive selection is selection for the trait; negative selection is against the trait.

tion. They acknowledge cultural choice as affecting style, but deny it the causal role assigned it in dual inheritance theories. Nor do they make the clear distinction between culture and behavior central to dual inheritance theory, which sees style as behavior.

According to these authors, since style is adaptively neutral, stylistic traits will display random changes through time, i.e., style will drift. If stylistic traits should go to fixation, that also will be caused by drift, or accident. The battleship curves of frequency seriation are seen as essentially the consequences of these random or Markovian processes. Conversely, functional changes through time will be the result of selective pressure. As drift is random, the resulting temporal patterns will be Markovian.

There are many problems with this approach. Traits can contribute to adaptiveness at one point, be neutral at another, and contribute, in a new way, to adaptiveness at yet another time. We will have no archaeological language to discuss these shifts if "style" is assigned solely to episodes of cultural drift. Further, as will be discussed below, the neutrality of a stylistic trait does not mean it will drift. Finally, this approach limits the potential range of questions to be asked about stylistic variation. I believe an approach to style closer to that outlined by Davis (1990) will ultimately be more fruitful to an evolutionary archaeology than that of Dunnell and others.

According to Davis (1990:19) "(a) *style* is a description of a polythetic set of similar but varying attributes in a group of artifacts, (b) the presence of which can only be explained by the history of the artifacts, (c) namely, common descent from an archaeologically identifiable artifact-production system in a particular state or states." Thus, style consists of attributes of artifacts (can be decoration, raw material, manner of execution, context of disposal, and so on) shared among a group of artifacts produced by a common production system. The idea of a production system is important here: a production system can have many steps, and selection can operate differently on different stages of the system. Production systems imply the "systemic knowledge" of Durham's definition of culture. A production system may be transmitted as a single meme, or as many memes. The memes may have few or many allomemes. Thus, dual inheritance theory asks the question, "what is being transmitted to produce this particular style?" The specification of a common history and shared descent is also central to an evolutionary understanding of style. The definition has at its core the historical connections required for evolutionary studies, and allows one to apply the concept of "descent with modification." The memes which contribute to the style are socially transmitted, and the expressed style is produced by the "surviving variants" (see above). We can study the cultural evolution of a style to determine what evolutionary forces have affected it through time without restricting the term to the result of only one force, natural selection (or the lack thereof). Rather, a style is a set of evolved, historical relationships reflected in the form of material culture.

The objects expressing a particular style are the product of behavior, part of the phenotype. Therefore, an expressed style reflects the interplay among cultural inheritance, genetic inheritance, and the environment (including social and "natural" elements). While the makers of the artifacts in question may have invested the objects they made with symbolic meanings, style itself does not provide a privileged window into the cultural system. The symbolic meaning of a style must be determined separately from the existence of the style. Meaning can be approached through hypotheses about the nature of transmission and the direction of selection acting on the style. Style cannot be directly linked to a "mental template," or platonic essences, nor need it necessarily reflect ethnicity or even vary with ethnicity. The relationship between a particular style and the ethnicity of its producers is an empirical question.

It is easy to imagine a pottery style, for example, which is the result of three memes: one for decorative motifs, one for vessel form, and the third for making the paste. It is also quite easy to imagine that the decorative style could be subject to one set of selective pressures, or to drift, and the paste making to completely different selective pressures. Cultural selection acting on allomemes for decorative motifs could result from ethnicity, while those on paste reflect pot function, irrespective of the ethnicity of the pots' users. Indeed, it should be possible, in principle, to establish the mode(s) of selection operating on particular elements of a style. For example, O'Brien and Holland see the effects of directional selection in the changes in the wall thickness of Woodland Period pottery from the south central United States to be reflecting cooking practices.

Directional selection is only one of the modes of selection. Directional (or directive [Mayr 1982:588]) selection occurs when, in Mayr's words, selection favors one tail of the curve of variation (imagine a bell-shaped curve), and discriminates against the other. The "direction" of selection pulls the mean of the curve toward the favored tail. Stabilizing selection occurs when selection is directed against both tails, "all deviations from the 'normal' are discriminated against" (Mayr 1982:587). Diversifying or disruptive selection occurs when selection favors both tails of the curve over the mean. Mayr also identifies a fourth mode, "catastrophic selection," when a trait at the tail end of the curve permits its carrier to survive some catastrophe which wipes out most of the bearer's conspecifics. Of course, such survival may be related to the trait, or to luck (Gould 1989).

This approach to style subsumes other recent definitions and usages of the term *style*. *Isochrestism*, for example, is Sackett's (1990 and citations therein) term for the constraints placed on formal variation of material cultural objects by ethnicity. He argues that "there normally exists a spectrum of equivalent alternatives...for attaining any given end in manufacturing and/or using material items, I refer to these options as constituting *isochrestic* variation" (1990:33). According to Sackett, isochrestic style is the product of the choices made among isochrestic variants by artisans based on their enculturation within their ethnic group. In the words of dual inheritance theory, there are two possible evolutionary forces at work to produce isochrestic variation: transmission and cultural selection. (1) Isochrestic variation could be produced when only certain memes or allomemes are socially transmitted among members of a reference group which identifies itself as an ethnic group, or (2) a range of allomemes are transmitted, but only a few are culturally selected because of values about what is ethnically appropriate. The first seems closest to Sackett's definition. Isochrestism is readily absorbed into a dual inheritance theory of style. Isochrestism would give stylistic evolution within a particular tradition an appearance of directionality by limiting variability even though, in Sackett's view, the style does not actively function as an ethnic marker, or have any function at all. Isochrestism channels variation in this case, mimicking the effects of selection.

Plog (1990:62), following Weissner (1985), recognizes *symbolic* and *iconologic* stylistic variation. Symbolic variation "presents information about similarities and differences that can help reproduce, alter, disrupt, or create social relationships." Symbolic variation is the result of how people learn how to make and decorate objects in a social context by comparing their work with that of others and "then imitate, differentiate, ignore, or in some way comment on how aspects of the makers or bearer relate to their own social and personal identities." This carries the implication, basic to much recent thought (papers and references in Conkey and Hastorf 1990), that style is actively negotiated. Iconological styles are a special subset of symbolic styles which have a specific audience.

From the standpoint of dual inheritance theory these distinctions reflect how cultural selection is operating on the style: i.e., the evaluations of the makers (and the audience) of particular stylistic variants against their secondary values about what material cultural forms are necessary or appropriate in certain broad or narrow social contexts. According to Durham, "cultural selection always involves some kind of comparative evaluation of variants according to their consequences" (1991:199). Crucial here, particularly with regards to the audience, is the mode of cultural selection; is it imposed or not?

NORTHWEST COAST ART

Northwest Coast art is, of course, one of the best-known art styles in the world. What we usually think of as Northwest Coast art is primarily that of groups living on the northern half of the coast in the 19th century. This northern style is marked by the use of form-lines and ovoids (Holm 1965) and balanced symmetry and motifs that interlock and flow into each other. In the southern style, motifs appear singly, they do not interlock, and the form-lines and ovoids are rare. The southern style is also more variable, and there may be less art on the southern coast (Borden 1983; Carlson 1983a,b; Suttles 1983, 1990; Ames, in press). The conventions governing the style were much stronger in the north. It is almost impossible to distinguish the tribe of origin for two-dimensional forms (e.g., boxes), though, to a practiced eye, it is somewhat easier for sculptural forms such as totem poles and masks. In the south regional differences can be quite marked, as between the Coast Salish and the Chinook.

Regionalism in the art style is paralleled by other cultural practices during the recent past, including the mutually exclusive spatial distributions of cranial deformation, and labret wear (the former in the north, the latter in the south). These regional patterns have changed across the last several millennia, as will be discussed below. First, I will describe how the current distribution of the regional styles has evolved.

Ninteenth century northern Northwest Coast art is well represented in the world's museums. Though creation of it almost ceased by the early years of this century, it has undergone an explosive revival since the late 1960s. This revival has been fueled, among others things, by:

1. Holm's analysis and codification of the formal rules governing the use of space and of the basic design elements which structure the art (Holm 1965);
2. A concurrent cultural revival of native American societies along the coast, and a renewal of some of the traditional social practices, like potlatches, to which the art was central;
3. A widespread appreciation and interest in the art among Euroamerican and Eurocanadians who see the art as symbolizing the region in which they live, and the use of the art as a marker of civic pride. For example, the logo of the Seahawks (the Seattle, Washington, football team) is in the Northwest Coast style. However, some southern variants of the style have virtually disappeared. In some areas, like the Lower Columbia River of Washington and Oregon, the local variant may be extinct. Thus, in the evolution of Northwest Coast art, one variant has almost become fixed. This situation could change if the other old variants are revived, or new ones develop. Among the reasons for the success of the Northern variant are the following:

    a. Frequency. In the 19th century, there appear to have been many more pieces of northern art produced. The reason why less art was produced on the southern coast is interesting (Suttles 1983) but need not detain us here. Thus, there were more objects to end up in museums and collections from the north.

b. Contact and disease history. The vast majority of materials were collected between 1875 and 1930 (Cole 1985), by which time "the city of Washington contained more Northwest Coast material than the state of Washington, and New York City probably housed more British Columbia material than British Columbia herself" (1985:286). Populations in the far southern Northwest Coast (coastal Washington and Oregon) were decimated and even destroyed by diseases by the early third of the 19th century, while those reproductively viable populations persisted to the north, producing art until the late 19th century. Thus, by the time serious collections started to be made, the art tradition in the far south may have ended or have been seriously disrupted. Increased production of the art may have been one of the directions resistance to European dominance took among northern groups (Cole and Chaikin 1990).

c. Research history. The focus of early anthropological research and collection on the Northwest Coast was with those still viable, functioning groups on the northern coast, reinforcing the effects of point b. Northern Northwest Coast excited enormous interest in the late 19th century, as it still does. This interest is one factor in the avidity in which the art was collected, and the availability of enormous collections of the northern art for study, and to shape the direction of the revival. In short, when the selective environment changed in the 1960s, and the revival of Native culture began, the northern style was the only variant available to them. The revival also began first among the northern groups who had originally produced the style. Thus, the recent evolution of Northwest Coast is the result of the interplay among intersocietal selection (contact with Europeans, diseases, the effects of European tastes and aesthetic judgments, missionization, and so forth), choice (resistance to contact), imposition (both by Europeans and by natives operating within the traditional prestige system and the need for art object for that system to work) and even chance (the effects of the frequency of art objects on the persistence of the northern style).

Northwest Coast art, or elements of it, may be among the oldest art styles (as the term has been defined here) still being produced in the world. It is possible to establish minimal ages for some elements of Northwest Coast art.

1. Joined nose and eyebrows in relief produced by carving away surrounding material (the creation of negative space [Holm 1965]). Eyes are also in relief. The earliest known example is an antler handle radiometrically dated to between 3000 and 1500 B.C.[2] was recovered from Glenrose Cannery, a site located near Vancouver, British Columbia (Matson 1976). A similar handle was recovered in Prince Rupert Harbor, on the northern coast of British Columbia, which postdates 500 B.C. The same raised brow and nose is present in a carved pumice figurine postdating A.D. 1500 recently recovered near Portland, Oregon. The design element is both geographically widespread and of great time depth.

2. Human and animal figures are commonly joined or intermingled. Antler spoons recovered from the Pender Island Site in southern British Columbia display zoomorphic elements joined at the mouth. Carlson (1991) has dated these spoons to 1650 BC. (The decoration of the ends of the spoons' handles, and other attributes of these spoons, such as the rendering of eyes also fall within the canons of 19th century Northwest Coast art.)

3. Bilateral symmetry and "visual punning." A common feature of the historic art is splitting an animal or human form down its middle and showing the two sides in profile, sometimes facing each other, sometimes facing away. Motifs were polyvalent in meaning, and several forms or elements were combined onto one. A sea mammal bone club displaying these characteristics from southern British Columbia has been AMS dated to 1895 (McMillan and Nelson 1989).

4. Some of the form-line conventions described by Holm for the 19th century art (Holm 1965), specifically the "split U-form" design, are present on an elaborately carved atl atl with a date of A.D. 210–440. This piece also has interlocked motifs, in this case a sea-monster(?) interlocked with a human face, as well symbolism which is readily connected to 19th century themes (Fladmark *et al.* 1987). Form-lines are also present on a red cedar handle recovered from the Lachane site (Gbto33) on the northern British Columbia coast, dating to 1630.

5. Materials recovered from wet sites such as Lachane and Hoko River (Inglis 1976; Croes 1988), as well as isolated pieces like the atl atl described above, clearly indicate that the carving skills and techniques of the 19th century carvers were in use at least 2000 years ago.

---

[2] All dates are calibrated dates using the University of Washington's Quaternary Laboratory radiocarbon calibration program version 2.1(**)

This list is not exhaustive, but makes the point. It is clear that Northwest Coast art constitutes a style, in the sense of Davis's definition, of great antiquity. However, the history of the regional variants is not known. Generally, the early individual examples of the art, like the spoons from Pender, seem suggestive of the northern style, whatever their provenience. On the other hand, these objects would not raise eyebrows in a collection of 19th century art from the south, either. Northwest Coast art has not been static over the last 4000 years. Some motifs have appeared and disappeared; some categories of decorated objects have appeared and disappeared (papers in Carlson 1983b). In the same way, labrets were originally worn along the entire coast by both sexes; historically the practice was limited to high-status women on the southern coast.

The **social** context within which cultural selection acted on the art has also evolved. The decorated spoons at the Pender Island site predate the earliest plank house village on the coast by some 800 years. Villages and towns of large, rectangular houses of western red cedar (*Thuja plicata*) planks and timbers were the dominate residential form along the coast at first contact with Europeans. These houses, along with being residences and food processing factories, were the stages on which potlatches and other ceremonies displaying the art were performed. Indeed, these spoons may even predate the heavy reliance on stored salmon on which the Northwest Coast prestige system is supposed to have depended. Croes and Hackenberger (1988) suggest storage of salmon did not become economically significant in the Coast Salish area of the southern Coast (where the spoons were recovered) until perhaps 2000 to 2500 years ago. Further, on the basis of present evidence, the art style predates the evolution of the Northwest Coast's system of social stratification.

Within the coevolutionary framework laid out in this paper, a number of problems arise with regard to Northwest Coast art, given the foregoing discussions. I do not have answers for these questions, but they suggest fertile and substantive research directions:

1. What are the memes and allomemes being transmitted from generation to generation? Can we discern changes in the information content of the memes? For example, Northwest Coast woodworking techniques are independent of the art style, which can be executed in stone, wood, bone, and now in gold and silver among other media. On the other hand, there are some motifs which seem to appear only in stone, for example. What this question requires are data on variation in form, media, and technique which are not currently available. It also requires information on the sampling problems which structure our current data base. An artifact sample of 18,000 tools recovered from nine sites in Prince Rupert Harbor, British Columbia, contains less than 300 objects of stone and bone with any decoration

at all. Why so few? Most decorated objects were wooden and so do not commonly preserve in archaeological sites. In addition, there is evidence that some classes of stone objects were disposed away from residential sites. A coevolutionary approach requires that we control as much of the potential sources of variation and error as possible.

2. What nonconveyance forces have been at work (e.g., innovation, synthesis, migration, diffusion, or cultural drift)? These may be the forces crucial to understanding the regional patterns of Northwest Coast art already described. This also raises issues about innovation, culture contact, and diffusion which have not been fashionable during the past three decades. It is in this context that issues of cultural drift become important, rather than in the single context of style versus function.

3. What conveyance forces have been acting on Northwest Coast art (transmission [Boyd and Richerson 1985], natural selection, cultural selection)? Boyd and Richerson's indirect bias mode of transmission provides a bases for constructing hypotheses about transmission effects. The art may have been an indicator trait (as it currently is) in the *indirect transmission* of crucial Northwest Coast cultural practices—the art may in fact have functioned as an indicator trait for the transmission of sets of cultural values, for other memes in other words.

In this case, as social evolution proceeded on the coast, and stratification evolved, the art continued to play its role in the transmission of socially important memes, but those memes changed. Carlson (1992) has argued that the art was originally shamanic, as much of it was in the 19th century (Jonaitas 1985). If this is so, the art was simply co-opted for the transmission of other memes when social stratification evolved, in a manner analogous to the co-option by natural selection of social interaction for the transmission of culture.

It is also within this problem that we address the question of the *locus* of cultural selection—what are the relevant reference groups? During the late prehistoric and early historic periods we can frame questions about the roles of title holders who commissioned pieces, the specialized carvers who executed them, and the rest of the population that witnessed the results. Indeed, a significant question becomes who produced the art? This is of course relevant to questions about relationships between social evolution and the organization of production (e.g full-time versus part-time specialists). But it also may be significant for the transmission of the art, and the nature of the culture selection acting on the art. A narrowly based guild of carvers could affect variation in a manner similar to Sackett's isochrestic style, and the differences between north and south reflect the number of specialized carvers. Another way to phrase that question is whether cultural selection acting on the art was through choice or imposition, and if imposition, was it from the titleholders, the artists, or the rest of society. This could be tested by examining the manner in which the art varied prior to the evolution of social stratification and/or craft specialization, and after their evolution.

It is also possible to suggest hypotheses about the regionalization of Northwest Coast art in the 18th and 19th centuries:

1. The regionalization of Northwest Coast art could be the result of stabilizing selection in which cultural selection against variation was weaker in the south than in the north (e.g., choice versus imposition).
2. The regionalization of Northwest Coast art could be the result of disruptive selection on a regional basis (selection favoring the evolution of regional styles) coupled with strong stabilizing selection on the northern variant. I have argued elsewhere that region was an important aspect of social personhood on the coast (head deformation versus labrets) and that the same selective forces affected the art (Ames in press).
3. The appearance of stabilizing selection is the result of isochrestism operating to limit variation: regional variation reflects the passive effects of ethnicity as expressed regionally.

I have not formalized or tested these hypotheses, because that is beyond the scope of this paper. However, the hypotheses could be formalized and tested using either the formal methods of Boyd and Richerson or the more qualitative methods of Durham. I have elsewhere presented a model to account for some of these patterns (Ames in press).

In any case, I have attempted here to show that dual inheritance theory provides a rich and productive approach to the issue of style which does not require the distinction between style and function developed by Dunnell (1978), and which subsumes many recent developments in archaeological theories of style. Dual inheritance theory is quite robust in this regard.

SUMMARY AND CONCLUSIONS

I have attempted to do two things in this paper: summarize Durham's theory of coevolution, and to show that an approach to style, based on Davis's (1990) definition, is more appropriate and useful to a Darwinian archaeology than that developed by Dunnell and others. Style was defined above as "a set of evolved, historical relationships reflected in the form of material culture." A key element to Davis's definition of style was that artifacts exhibiting a common style do so because they either share a common production system, or productions systems which are historically linked to a common ancestral production system. Production systems may be encoded in one or more memes (and allomemes) and therefore cultural selection can act differentially on different portions of a single production system.

This approach to style is appropriate to a Darwinian archaeology because it permits one to pursue issues of historical relationships, as well as function and adaptability in a flexible and productive manner. It does not limit "style" to episodes of drift, which, within coevolution, is only one of several nonconveyance forces.

Many of the research questions outlined in the discussion on Northwest Coast art could have been developed within other theoretical frameworks—which suggests that dual inheritance theory is compatible with some other, non-Darwinian anthropological theories. However, in conclusion I would suggest that coevolution has three general advantages over other available theories in Anthropology. First, no other currently accessible non-Darwinian theory allows one to deal with the cultural evolution of Northwest Coast art—and similar matters—and the evolution of the biological foundations of language, cognition, and culture within a single framework. I believe Darwinian approaches generally, and coevolution in particular, will prove extremely powerful in providing answers about the evolution of hominid culture. Second, of available Darwinian approaches, coevolution seems both the richest, and the most likely to be able to absorb and utilize anthropological insights available from other paradigms (e.g., Mithen 1989). It is therefore more likely to be broadly adopted and used, is more compatible with currently available data sets, and does not require abandoning the central core of anthropological and archaeological knowledge about culture.

Third, with Shennan (1989), I believe that coevolutionary theory will eventually allow us to theoretically link our everyday lives with broad-scale social and cultural changes. Shennan has recently suggested that dual inheritance theories provide actual mechanisms by which our everyday practices, the *habitus* (Bourdieu 1977; Giddens 1984), can actually reproduce social structure. He emphasizes Boyd and Richerson's (1985) biased transmission. I add to this Durham's "imposed selection" and power. In this paper, I suggested that the evolution of the Roman Empire was at least in part the result of the *habitus* of the Roman oligarchy. The foregoing discussions carry the implication that our concept of large-scale social processes may, in part, be an issue of our "monitoring position": one reference group's "habitus" may be someone else's catastrophic selection. How events will appear to us will be determined by which reference group provides us our vantage point. Are the evolutionary forces of coevolution sufficient to generate the large-scale events of social evolution? If so, then we can abandon the search for the laws of culture at the heart of processual archaeology, and the laws of history at the heart of historical materialism.

What remains a central and exciting challenge then is finding the causes of broad-scale similarities in social evolution (i.e., Johnson and Earle 1987). Braun (1991) has already hazarded an explanation, but it, like all of the foregoing, is preliminary.

REFERENCES

Ames, K.M., in press, Art and Regional Interaction Among Affluent Foragers on the North Pacific Rim, in: *Development of Hunting-Gathering-Fishing Maritime Societies on the Pacific* (A.B. Onat, ed.), Washington State University Press, Pullman.
Bettinger, R.L., 1990, *Hunter-Gatherers: Archaeological and Evolutionary Theory*, Plenum Press, New York.
Binford, L.R., 1962, Anthropology as Archaeology, *American Antiquity* 2(2):217–228.
Borden, C.C., 1983, Prehistoric Art in the Lower Fraser Region, in: *Indian Art Traditions of the Northwest Coast* (R.L. Carlson, ed.), Simon Fraser University Press, Burnaby, pp. 131–166.
Bourdieu, P., 1977, *Outline of a Theory of Practice*, Cambridge University Press, Cambridge.
Boyd, R., and Richerson, P.J., 1985, *Culture and the Evolutionary Process*, University of Chicago Press, Chicago.
Braun, D.P., 1991, Are There Cross-cultural Regularities in Tribal Social Practices? in: *Between Bands and States: Sedentism, Subsistence and Interaction in Small Scale Societies* (S. Gregg, ed.), Occasional Paper No. 9, Center for Archaeological Investigations, Southern Illinois University, Carbondale, pp. 423–444.
Carlson, R.L., 1983a, Prehistory of the Northwest Coast, in: *Indian Art Traditions of the Northwest Coast* (R.L. Carlson, ed.), Simon Fraser University Press, Burnaby, pp. 13–32.
Carlson, R.L., 1983b, *Indian Art Traditions of the Northwest Coast*, Simon Fraser University Press, Burnaby.
Carlson, R.L., 1991, The Northwest Coast Before A.D. 1600, in: *Proceedings of the Great Ocean Conferences*, Volume 1, The Oregon Historical Society, Portland, pp. 109–137.
Carlson, R.L., 1992, Paleo-Shamanism on the Northwest Coast, Paper presented at the 45th Annual Northwest Conference, Burnaby.
Cole, D., 1985, *Captured Heritage, The Scramble for Northwest Coast Artifacts*, University of Washington Press, Seattle.
Cole, D., and Chaikin, I., 1990, *An Iron Hand Upon the People: The Law Against the Potlatch on the Northwest Coast*, University of Washington Press, Seattle.
Conkey, M.W., and Hastorf, C.A., 1990, *The Uses of Style in Archaeology*, Cambridge University Press, Cambridge.
Croes, D.R., 1988, The Significance of the 3000 B.P. Hoko River Waterlogged Fishing Camp in our Understanding of Southern Northwest Coast Cultural Evolution, in: *Wet Site Archaeology* (B.A. Purdy, ed.), The Tellford Press, Caldwell, pp. 131–152.
Croes, D.R., and Hackenberger, S., 1988, Hoko River Archaeological Complex: Modeling Prehistoric Northwest Coast Economic Evolution, *Research in Economic Anthropology* Supplement 3:19–86.
Davis, W., 1990, Style and History in Art History, in: *The Uses of Style in Archaeology* (M.W. Conkey and C.A. Hastorf, eds.), Cambridge University Press, Cambridge, pp 18–31.

Dunnell, R.C., 1978, Style and Function: A Fundamental Dichotomy, *American Antiquity* 43:192–202.

Dunnell, R.C., 1980, Evolutionary Theory and Archaeology, in: *Advances in Archaeological Method and Theory*, Vol. 6, (M.B. Schiffer, ed.), Academic Press, New York, pp. 35–99.

Dunnell, R.C., 1992, What is it that Actually Evolves?, Paper Presented at the 57th Annual Meetings of the Society for American Archaeology, Pittsburgh.

Durham, W.H., 1991, *Coevolution: Genes, Culture, and Human Diversity*, Stanford University Press, Stanford.

Eldredge, N., 1986, Information, Economics and Evolution, *Annual Review of Ecology and Systematics* 17:351–369.

Fladmark, K.R., Nelson, D.E., Brown, T.A., Vogel, J.S., and Southen, J.R., 1987, AMS Dating of Two Wooden Artifacts from the Northwest Coast, *Canadian Journal of Archaeology* 11:1–12.

Gellner, E., 1988, *Plough, Sword and Book: The Structure of Human History*, University of Chicago Press, Chicago.

Giddens, A., 1984, *The Constitution of Society, Outline of the Theory of Structuration*, University of California Press, Berkeley.

Gould, S.J., 1989, *Wonderful Life: The Burgess Shale and the Nature of History*, Norton, New York.

Hamilton, W.D., 1964, The Genetical Evolution of Social Behavior: I and II, *Journal of Theoretical Biology* 7:1–52.

Holm, B., 1965, *Northwest Coast Indian Art: An Analysis of Form*, Monograph No. 1, Thomas Burke Memorial Museum, University of Washington Press, Seattle.

Inglis, R., 1976, 'Wet' Site Distribution—The Northern Case GbTo 33—The Lachane Site, in: *The Excavation of Water-Saturated Archaeological Sites (Wet Sites) on the Northwest Coast of North America*, (D. Croes, ed.), Archaeological Survey of Canada Paper No. 50. National Museums of Canada Mercury Series, Ottawa, pp. 71–108.

Johnson, A.W., and Earle, T., 1987, *The Evolution of Human Societies from Foraging Group to Agrarian State*, Stanford University Press, Stanford.

Jonaitas, A., 1985, *Art of the Northern Tlingit*, University of Washington Press, Seattle.

Keesing, R., 1974, Theories of Culture, *Annual Review of Anthropology* 3:73–97.

Lukes, S., 1974, *Power: A Radical View*, Macmillan and Co., London.

McMillan, A.D., and Nelson, D.E., 1989, Visual Punning and the Whales Tail: AMS Dating of a Marpole-age Art Object, *Canadian Journal of Archaeology* 13:212–218.

Matson, R.G., 1976, *The Glenrose Cannery Site*, Archaeological Survey of Canada Paper No. 52. National Museums of Canada Mercury Series, Ottawa.

Mayr, E., 1982, *The Growth of Biological Thought: Diversity, Evolution, and Inheritance*, Harvard University Press, Cambridge, MA.

Millett, M., 1990, *The Romanization of Britain, An Essay in Archaeological Interpretation*, Cambridge University Press, Cambridge.

Mithen, S.J., 1989, Evolutionary Theory and Post-Processual Archaeology, *Antiquity* 63:483–494.

O'Brien, M.J., and Holland, T.D., 1990, Variation, Selection, and the Archaeological Record, in: *Advances in Archaeological Method and Theory*, Vol. 2 (M.B. Schiffer, ed.), University of Arizona Press, Tucson, pp. 31–79.

O'Brien, M.J., and Holland, T.D., 1992, The Role of Adaptation in Archaeological Explanation, *American Antiquity* 57(1):36–59

Plog, S., 1990, Sociopolitical Implications of Stylistic Variation in the American Southwest, in: *The Uses of Style in Archaeology* (M.W. Conkey and C.A. Hastorf, eds.), Cambridge University Press, Cambridge, pp 61–70.

Rindos, D., 1984, *The Origins of Agriculture: An Evolutionary Perspective*, Academic Press, New York.

Rindos, D., 1986, The Evolution of the Capacity for Culture: Sociobiology, Structuralism, and Cultural Selection, *Current Anthropology* 27:315–332.
Sackett, J.R., 1990, Style and Ethnicity in Archaeology: The Case for Isochrestism, in: *The Uses of Style in Archaeology* (M.W. Conkey and C.A. Hastorf, eds.), Cambridge University Press, Cambridge, pp. 32–43.
Shennan, S., 1989, Cultural Transmission and Cultural Change, in: *What's New? A Closer Look at the Process of Innovation* (S.E. van der Leeuw and R. Torrence, eds.), Unwin Hyman Ltd, London, pp. 330–346.
Suttles, W., 1983, Productivity and its Constraints: A Coast Salish Case, in: *Indian Art Traditions of the Northwest Coast* (R.L. Carlson, ed.), Simon Fraser University Press, Burnaby, pp. 67–88.
Suttles, W. (ed.), 1990, *Handbook of North American Indians*, Vol. 7, Smithsonian Institution, Washington, DC.
Weissner, P., 1985, Style or Isochrestic Variation? A Reply to Sackett, *American Antiquity* 50(1):160–166.
Wolf, E.R., 1990, Distinguished Lecture: Facing Power—Old Insights, New Questions, *American Anthropologist* 92:586–596.
Yoffee, N., 1991, Maya Elite Interaction: Through a Glass, Sideways, in: *Classic Maya Political History: Hieroglyphic and Archaeological Evidence* (T. P. Culbert, ed.), Cambridge University Press, Cambridge, pp. 285–310.

Chapter 8

Style, Function, and Cultural Evolutionary Processes

ROBERT L. BETTINGER, ROBERT BOYD, AND
PETER J. RICHERSON

INTRODUCTION

When explaining human behavior, anthropologists frequently distinguish the things that people do of their own free will from the things they do because they have to. In much of anthropology, and most American archaeology, this is the difference between style and function. Functional behaviors are the things people are constrained to do; stylistic behaviors are the things people do when unconstrained. Where necessity stops and free choice begins is, of course, a classic problem of social science theory, but wherever the boundary is placed, it is generally implied that the domains thus divided are not of equal importance (Bettinger 1991:49–50). Few straddle this fence: Materialists emphasize function and downplay style; structuralists and postmodernists do the opposite. Recent attempts to apply neo-Darwinian concepts to the archaeological record predictably side with the materialist tradition, repeating the premise that it is most important to explain functional behavior; stylistic behavior is interesting only for localizing social units in time and space.

ROBERT L. BETTINGER • Department of Anthropology, University of California, Davis, California 95616. **ROBERT BOYD** • Department of Anthropology, University of California, Los Angeles, California 90024. **PETER J. RICHERSON** • Division of Environmental Studies, University of California, Davis, California 95616.

Any attempt to create a rigid boundary between style and function will fail. For example, the attempt to use free will as a distinction founders on the fact that conforming to stylistic conventions of speech, dress, and belief is frequently compulsory and almost always sanctioned. We may often have more opportunities for free choice of mundane utilitarian objects. Style often has functions, and the most basic functions—eating, defecating, and having sex—are usually done in style.

In this paper we argue that materialists forfeit too much when they dismiss the importance of style for humans. Human stylistic behavior over the last 40,000 years is extraordinarily extensive and elaborate. This is critical because for neo-Darwinians, complex, richly structured forms always signal the operation of natural selection or related evolutionary forces. There are simply no known material processes except natural selection, and analogous evolutionary forces in the cultural realm capable of accounting for phenomena that appear to be "designed" (see Dawkins 1986 for an excellent introductory summary of the adaptationist form of this argument). From this view, art objects, languages, and supernatural ideologies seem as much to be the product of evolutionary processes as subsistence technology and cannot be ignored.

With the advent of unambiguous stylistic features in the archaeological record at the Upper Paleolithic Transition, subsistence strategies also improved and populations of humans jumped in size (Stringer and Gamble 1993). We argue that this coincidence is not accidental. Style has functions. The style–function dichotomy embraced by materialists and nonmaterialists alike obscures understanding of the fundamental processes that generate human behavior.

Treating style and function as a dichotomy arises from an oversimplified picture of evolutionary processes. In animals, style arises via sexual selection and perhaps more generally via social selection (West-Eberhardt 1983). Evolutionary biologists since Darwin have engaged in a complex debate about the functionality of plant and animal style. However, this debate has hardly ever had the character of dividing sexually selected traits off as inconsequential. The modern debate has focused on whether style is counterfunctional (the runaway hypothesis) or whether it is an index of the overall fitness of a potential mate (the handicap hypothesis). Too much time and effort go into style for it to be neutral!

Human cultural styles cannot be explained without understanding the cultural analogues of the sexual selection mechanism. The conventional style–function dichotomy is a result of not taking account of the variety of these forces in cultural systems, and how they relate to the action of natural selection of adaptations. Human culture is influenced by a complex of evolutionary forces that ultimately derive from the operation of natural selection, but which have proximal properties that differ substantially from it (Campbell 1965; Boyd and Richerson 1985). Several of these involve the choices of cultural mates and parents and are like mate choice sexual selection. Others are perhaps more direct stand-ins for natural selection through psychological predispositions (Cosmides and Tooby 1992). When such "related evolutionary forces" are taken into account, the rigid distinction between style and function dissolves. Stylistic variation responds to a complex of random and directional evolutionary forces and can serve important functions precisely because it is arbitrary and symbolic. Explicit evolutionary models of stylistic variation clarify our understanding of style and its relation to function and culture history, and improves our understanding of the patterns style might leave in the archaeological record.

In the first part of this paper, we review recent debates on the implications of stylistic behavior for archaeology. Then we analyze the limitations inherent in assuming that evolutionary processes can be collapsed into selection acting on functional attributes and random effects acting on stylistic variation. Finally, we argue that recent advances in the theory of cultural evolution provide a reasonable account of the processes that affect the complexity of style and function in culture.

STYLE AND FUNCTION IN NEO-EVOLUTIONARY PERSPECTIVE

Style versus Function

> Natural selection is the primary explanatory mechanism in scientific evolution....Style and function are defined in terms of natural selection. Because of the distributional entailments of natural selection, each has a distinctive, wholly predictable distribution in the archaeological record. (Dunnell 1980:49,88)

> Style denotes those forms that do not have detectable selective values. Function is manifest as those forms that directly affect the Darwinian fitness of the populations in which they occur...The dichotomy is mutually exclusive and exhaustive. (Dunnell 1978:199)

> Stylistic variation is selectively neutral. Hence stylistic traits are sorted stochastically: by drift. (Neiman 1993:1)

The position of Dunnell (1978, 1980), a prominent archaeological exponent of the rigid style–function dichotomy, serves as an excellent starting point for our counter-argument. He argues that the distinction between style and function in anthropology is essentially between behaviors that are subject to processual explanation and behaviors that are not. Because Dunnell advocates a neo-Darwinian view of process, for him *functional* refers to things explicable as adaptations related to natural selection and *style* means, effectively, afunctional or neutral—things without direct positive or negative selective value.

Dunnell argues that because they are free of selective constraint, stylistic traits will vary stochastically, much like adaptively neutral traits in biology. He notes that these properties make style especially appropriate as measures of time (e.g., in seriation), social interaction, and culture history (cf. Neiman 1993). As with other techniques that employ presumably neutral traits to measure descent relationships (e.g., noncoding DNA resemblances, lexicostatistics), shared features of style are taken to be homologous similarities reflecting common cultural heritage. Dunnell observes one complication in this simple style–function distinction: if a functional requirement admits alternative solutions, the same trait can be stylistic (neutral) and functional, depending on the level (scale) at which it is defined. Variants of a functional trait (e.g., Z-twist and S-twist cordage) may be neutral with respect to each other, even though all are utilitarian.

Dunnell's definition of style and function is widely accepted by those interested in applying evolutionary principles to the archaeological record. Most materialist archaeologists agree that functional features will be nonrandomly patterned as a result of selection, and features of style will be merely stochastic. This is frequently read as meaning that the former are subject to processual explanation, the latter are not (e.g., Kirch 1980; Leonard and Jones 1987; O'Brien and Holland 1992; Neiman 1993).

In portraying art and style—the things anthropologists have historically identified as distinctively "cultural"—as beyond the reach of neo-Darwinian explanation, Dunnell articulates a traditional tenet of materialist anthropological inquiry, expressed first in modern form by Steward (1938), and subsequently in increasingly extreme form, by early cultural materialists (Harris 1968), neofunctionalists (Vayda and Rappaport 1967), New Archaeologists (e.g., Binford 1962; cf. White 1959), and, most recently, human evolutionary ecologists (e.g., O'Connell et al. 1983). Dunnell simply operationalizes the traditional argument through the prediction that art and style always pattern randomly.

Style as Style

> Every style is necessarily prelimited.... The range of its channeled skills will extend so far; beyond they fail. Then we say that the style has exhausted itself, its characteristic pattern has broken down.... It is commonplace that all aesthetic styles, rise and fall and perish. (Kroeber 1948:329–330)

Many anthropologists, of course, dispute the materialist account of style and the nature of stylistic change. Kroeber (1948) was one of many for whom stylistic change was directed rather than random. Kroeber viewed styles as basic themes (analogous to styles in art or music) on which cultures elaborated. Because it seemed inconceivable to him that such elaboration could continue beyond a climax in which the possibilities inherent in the style were exhausted, Kroeber believed that stylistic change followed a nonrandom historical trajectory, a position that Dunnell and others explicitly reject. Claims of this kind

(Sahlins 1976) are often presented as alternatives to the functional account style, but this is not necessary. Even among Kroeber's lengthy ruminations, one can find the kernel of an idea reconciling the view that style shapes and constrains cultural change with the functionalist view that culture is adaptive. "...For things to be done well they must be done definitely, and definite results can be achieved only through some specific method, technique, manner, or plan of operations. Such a particular method or manner is called a style in all the arts....A style...may be said to be a way of achieving definiteness and effectiveness in human relations by choosing or evolving one line of procedure out of several possible ones and sticking to it" (Kroeber 1948:329).

Sackett (1982, 1985) has more recently labeled such behavior isochrestic: patterned behavior reflecting essentially arbitrary choices between essentially functionally equivalent ways of doing things. This is Dunnell's scale effect—specific variability in traits that equally satisfy the same adaptive function. Dunnell, Sackett, and Kroeber, then, all seem to agree that just how one skins a cat can be functionally less important than the fact that one skins it at all. They disagree fundamentally, however, in what this implies about the mechanisms driving the historical trajectories of alternate variants of cat-skinning.

For Kroeber (as exemplified by the latter quote) and Sackett, isochrestic variation is mainly a product of the formalization and routinization of technique, which makes the transmission of the knowledge about how to make a complex object easy to imitate, remember, and execute. These benefits evidently resulted when simpler forms of individual and social learning were replaced by arbitrary conventions that streamlined acquisition by cultural transmission and coordinated complicated cultural behaviors. Once craftsmen become skilled at making and using a tool one way, they may rationally resist change because of learning costs. This is consistent with formal models suggesting it pays to retain a suboptimal tool when searching for the optimal alternative is costly or error prone (Simon 1959; Heiner 1983; Boyd and Richerson 1992a). At the same time, as Kroeber and Richardson's (1940) classic paper on dress style shows, there is nothing in the concept of isochrestic variation that denies the possibility of nonrandom historical trajectories of change.

Others working with style find isochrestic variation methodologically problematic and favor stylistic inquiries that emphasize iconic and symbolic variation. Binford (1989:52–53) sees the style–function dichotomy as an opposition between conscious, explicitly rational, problem-solving behavior, on the one hand, and unconscious, rote-learned motor habits and socially or symbolically motivated behavior, on the other. Within the latter, he evidently now follows Weissner (1985:162) in equating isochrestic variation with the unconscious or rote-learned motor habits (Binford 1989:56, 58). Because it is always possible that what appear to be isochrestic variants connected with individuals or ethnic units actually have functional significance, Binford believes that when defining actors or actor-groups in the archaeological record, it is safest to focus on the most obvious sorts of nonfunctional variation related to social or ideological behavior. The closer a style is to purely symbolic, the less likely it is to be functional, i.e., patterned by rational choice. Backed and self bows are at some level functionally equivalent but this equivalence is not guaranteed so generally as would be the functional equivalence of alternative geometric designs painted on them.

Binford and Weissner have discussed strategies for working with formal variation in material culture that consciously transmits information about social or personal identity. The emphasis here is on the use of the variation as emblems or icons of social or political groups (emblematic and iconological style) or expressions of individual identity within such groups (assertive style). In these cases, variation is said to be purely symbolic but serves a function (communication) and is surely not random. Weissner (1985:162); and evidently Binford (1989:54–55) argue that because social or ideological stylistic variation is manipulated to suit changing social and individual contexts, it should vary substantially through time and space, in contrast to isochrestic variants resulting from streamlined cultural decision-making, which are stable once established. As just noted, however, the isochrestic concept does not require this and, as Dunnell argues, isochrestic choices made by individuals can certainly give rise to behavioral change at both the individual and population levels.

In the main, Dunnell dismisses the relevance of these distinctions for the archaeological record. In contrast to Sackett, for whom cultural transmission streamlines (hence constrains) decision-making, for Dunnell cultural transmission is adaptive for the opposite reason: it broadens access to behavioral alternatives. It increases the amount of functionally significant variation from which individuals can choose and, thus, on which selection can act. This increases the speed and range of adaptive responses relative to simple genetic transmission, where population variation is more finitely constrained by such things as generation length, mutation rates, and existing genetic variation (Dunnell 1978:198). Selection sorts (hence patterns) the functional traits, leaving stylistic traits, symbolic and isochrestic, to drift randomly. If this is so, the methodological and ontological complexities of nonsymbolic isochrestic variation that charge the theoretical debate between Sackett, Binford, and Weissner are empirically unimportant. Style and function are more clearly distinct under this assumption, and nonstylistic variation is always functionally significant.

To summarize, if our review of the literature above is correct, archaeologists identify three types of artifact variation: (1) functional variation uncomplicated by communication of any sort, (2) functional variation preserved by rote social learning in which variants are qualitatively distinct but broadly equivalent in function, and (3) variation in iconic and symbolic traits that are arbitrary, functionless decorative elements. These three types seem to be

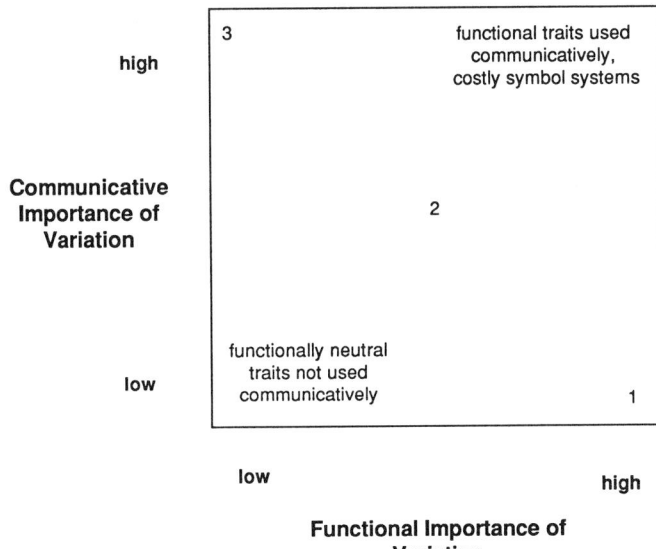

Figure 1. The three "types" described in the text may result from a tendency for actual variation to cluster along the diagonal as indicated by the numbers corresponding to the numbered items in the text. However, it is an empirical matter how thickly cases are scattered off the diagonal.

viewed as points in a two-dimensional space with one dimension representing function, ranging from completely functional to completely isochrestic, and the other dimension representing communication, ranging from variation that is highly salient as expressive or emblematic communication to variation that communicates nothing and is socially irrelevant. Note that although there might be some tendency for specific traits to lie on a diagonal line in Figure 1, nothing prevents the existence of off-diagonal cases, as Weissner, Binford, and Sackett repeatedly note. The upper right of Figure 1, for example, would accommodate the many known cases in which stylistic display is costly (e.g., Cohen 1974) or in which functional differences are meaningful as expressive or emblematic symbols, as when pastoralists take pride in owning cattle and despise their livestock-poor farming neighbors. By contrast, the lower left of Figure 1 would include cases where functionally neutral variation is completely ignored for communication, as Sackett supposes for variation in San projectile points.

There is little agreement about what Figure 1 means for the patterns one might expect to find in the archaeological record. Many commentators, represented by Dunnell, would apparently be comfortable arguing that variation projecting on the function dimension will be controlled by selection, while that projecting on the communicative dimension by random processes. Structuralists, represented in our brief review by Kroeber, imagine nonrandom processes, but emphatically not selection, to be acting on variation with high communicative function, perhaps leaving only variation near the bottom left of Figure 1 to random processes. In contrast, Weissner (1985:162) and Binford (1989:54–55) argue that variants in the bottom left of Figure 1 should be highly stable and not subject to random processes, a pattern Binford (1989:54) extends to many highly adaptive, nonsymbolic functional characters (i.e., variation in the lower right of Figure 1). Sackett (1986:630–631) seemingly rejects the notion that isochrestic variation will consistently conform to any specific pattern.

A CRITIQUE OF THE SELECTIONIST PROGRAM IN MODERN ARCHAEOLOGY

We believe that one cannot operationalize the models of style presented by Dunnell, Kroeber, Sackett, Binford, and Weissner, much less differentiate them with respect to pattern, because nowhere in their writings can one find an explicit model of the cultural transmission and "selection" processes that give rise to stylistic variation. Opler (1964) took Kroeber to task severely on this count but the others mentioned here are equally culpable. Indeed, given the importance contemporary archaeology places on model-building it is remarkable that these individuals have not been more severely criticized on this point by processualists (postprocessualists, of course, have not overlooked the problem, but their critique is beyond the scope of this discussion). The absence of criticism is symptomatic of a tendency of contemporary materialism to reduce cultural process to selection/adaptation, with the implication that these are a clearly understood, straightforward processes. When this simple selectionist account fails, materialists are much too ready to abandon the inquiry (e.g., Binford and Weissner) or resort to explanation in terms of random factors (e.g., Dunnell).

The problem, in short, is that the belief that cultural variation is adaptive is not supplemented by a concern with the details of the processes through which adaptations actually arise. Despite references to the importance of cultural transmission (e.g., Dunnell 1978:198), the adaptationists are suspicious of models of those processes. The groundwork for this view was laid by White and subsequently explicated in detail by Binford and others in the New Archaeology movement. They exhorted archaeologists to abandon models of culture as a system of inheritance and replace them with models of culture as adaptation.

> A normative theorist is one who sees as his field of study the ideational basis for varying ways of human life.... For adherents of the normative school, the assumptions about units or natural "packages" in which culture occurs are dependent upon as-

sumptions about the dynamics of ideational transmission.... The normative view leaves the archaeologist in the position of considering himself a culture historian and/or paleopsychologist (for which most archaeologists are poorly trained)...[therefore] a new systematics, one based on a different concept of culture, is needed to deal adequately with the explanation of culture process...[that being]...culture as man's extrasomatic means of adaptation [White 1959:8]. (Binford 1965:203, 204, 205).

Dunnell, Sackett, and Binford (and of course many others) evidently believe that ignoring the details of cultural transmission is justified because selection favors faithful cultural reproduction.

If we acknowledge that a cultural system is a system of extrasomatic transmission for behaviorally relevant information from one generation to the next, then a cultural tradition in its reproductive mode would be most effective if the transmission of information from one generation to the next is exact and unchanged in the process. (Binford 1983:222)

I think it should be clear that discussing a cultural system in terms of...the dynamics of cultural reproduction...is not likely to help us understand the dynamics of descent with modification. (Binford 1983:222 [but compare Binford 1983:221])

For Binford (and most other adaptationists), White's definition of culture as an extrasomatic means of adaptation selectively favors exact or near-exact cultural transmission: the initial phase of each new cultural generation is a near-duplicate of the terminal adult phase of the preceding generation, differing only in minor and random ways, as Dunnell assumes. These authors believe that such faithful transmission renders transmission inconsequential. To predict human behavior in a particular environment, one need only determine what behaviors are adaptive in that environment—selection will sort things out so that explicit attention to the process of adaptation is unnecessary.

Experience in evolutionary biology suggests that this view is almost certainly wrong. Adaptationist thinking has been extremely useful to evolutionary biology, and adaptationists have offered many useful insights about the behavior of humans and other species. However, complete dependence on adaptionist thinking, particularly the simple version used by Dunnell, Binford, and others, forfeits the most powerful elements of Darwinian thinking because, contrary to the view widespread in anthropology, there is more to Darwinism than natural selection. Selection is just one of several Darwinian processes, and "natural selection" itself is a heterogeneous complex of processes, some of which do not produce adaptations in any intuitive sense.

More fundamentally, Darwinism is not a list of processes ordered by relative importance; it is a methodology guided by the central assumption that the key to understanding evolution is good bookkeeping. Even the simplest evolutionary forces interact at several levels in complex ways, and to understand these complexities requires concerted attention to accounting for how some inherited variation responds at the population level to forces like selection. It is the interaction of the forces, at various levels, that is of interest.

Sexual selection referred to already illustrates the importance of keeping careful accounts of the interaction of multiple processes. Darwin believed that exaggerated, presumably maladaptive, male characters like peacock tails arose because females preferred males with such characters. Darwin could not explain, however, why females prefer such males. This question is now being hotly debated by evolutionary biologists (Eberhard 1986; Pomiankowski 1988; Kirkpatrick 1989; Barton and Turelli 1991; Maynard Smith 1991). The runaway hypothesis, one of the competing positions in this debate, will illustrate our point. In this view, female preference for showy males is a pathological consequence of female choice (e.g., Lande 1981). Suppose there are cryptic males with practical camouflage tails and showy males with bright predator-attracting tails. If females who prefer males with showy tails are sufficiently common, their choices can increase the frequency of "showy-tail genes" even though such tails are otherwise disadvantageous. Such choices will also cause the genes that generate a preference for showy tails to co-occur with the genes that cause showy tails. As a result, an increase in the frequency of "showy-tail" genes will also cause an increase in the genes that cause females to prefer showy males. During the next generation, sexual selection will favor showy males even more strongly, which in turn will further increase the frequency of females who prefer showy males, further increasing the strength of sexual selection, and so on, until males become spectacularly elaborated. A key problem for the runaway hypothesis is the magnitude of the association between genes for tails and genes for female preferences, which in turn depends on the interaction between natural selection acting on both sets of genes, sexual selection, and the mechanics of linkage and recombination. It is simply impossible to understand this plausible evolutionary mechanism without detailed models that carefully track the net effects of this complex of interacting processes (Barton and Turelli 1991). Nor is this a singular case, similar complexities are confronted in the sexual selection debate (e.g., Hamilton and Zuk 1982; Ryan et al. 1990), models of speciation, models of the evolution of sex, recombination, mating systems, the shifting balance theory, and a number of other current problems in evolutionary biology.

The lesson is clear. The evolutionary interpretation of human behavior, contemporary and extinct, requires anthropologists to construct explicit models of cultural processes and calculate the implications of those models. Binford has relentlessly exposed the fallacies of interpretation that result when we try to intuit the meaning of archaeo-faunas without formal models that force us to keep track of various formation processes acting at various levels. Darwin's methods of "population thinking" encourage a similar attention to the details of how particular variants increase or fail to do so under the impact of specific environmental and social effects. It is quite clear that an evolutionary perspective of culture process requires models of cultural transmission that are analytically separate from models of other processes that act on that form of variation, such as selection.

In this regard, contemporary adaptationists in archaeology tend to follow the interpretive tradition of anthropology, which emphasizes generalizations about consequences, rather than the more process-oriented tradition of evolutionary biology (Bettinger 1991). The data of a particular archaeological case are "explained" by means of empirical generalizations about the archaeological record and by arguments about the larger "meaning" of those records. For contemporary adaptationists, the archaeological record implies the overwhelming importance of natural selection. Thus, adaptation, like progress for earlier scholars, is used as the interpretive tool to dissect and explain a case at hand.

In contrast, evolutionary biology devotes much of its effort to studying the actual processes of evolution. Genetics, population genetics, and population ecology are mostly about the processual inner mechanics of the inheritance of variation and its modification by the population-level impact of environment. The adaptive interpretation of the structure and behavior of particular organisms depends on the knowledge we have about these processes, gained from many kinds of studies of many kinds of species. Sometimes adaptive interpretations are fairly obvious and do not depend crucially on a close knowledge of process, but the opposite is quite often true. In the case of sexual selection, for example, the debate is tightly focused on the details of models of the sexual selection process and on the interpretation of data (e.g., large scale surveys of bird coloration by Hamilton and Zuk [1982], and of insect intromittent organs by Eberhardt [1986]). Rather than use the theory to interpret cases, cases are used to decide how the theoretical models apply. Only if this search leads to general conclusions do we obtain some warrant for a more general interpretive strategy.

PROCESSES OF STYLISTIC EVOLUTION

The traditional definition of style requires that behavioral variants not be subject to natural selection. Stylistic variants must be neutral with respect to natural selection (and selection-derived, adaptation-generating, decision-making effects such as Boyd and Richerson [1985] discuss under the headings of bias and guided variation). This definition fails to do justice to stylistic variation in three major ways. First, many isochrestic variants of utilitarian artifacts may be subject to frequency-dependent effects. When Qwerty keyboards are common, it is sensible to adopt them, even though rare keyboards (like

Dvorak) are actually better. Selection itself can maintain stylistic heterogeneity. Second, purely symbolic characters will come to have fitness effects if they become the object of choice, as in sexual selection. Third, stylistic variation may be controlled by evolutionary forces that generate nonrandom patterns, even in the neutral case. (It is perhaps also worth mentioning that natural selection in a randomly varying environment will tend to impose that randomness on functional variation under its control and that chaotic dynamics might mimic random variation.)

We sketch below a taxonomy of the processes that might affect the evolution of stylistic features. This discussion leads to two conclusions: First, well-defined cultural evolutionary processes can result in detectable, nonrandom patterns in adaptively neutral stylistic variation that will often be difficult to distinguish from the kinds of patterns that result from natural selection acting to produce adaptations. Second, some of the reasons for pattern in style have to do with indirectly functional features. The argument that there should be a simple distinction between random stylistic and adaptive functional patterning is supported neither methodologically nor ontologically.

Pattern Generated by Nonselective Random Processes

If individuals acquire stylistic traits by faithfully copying others, and then make innovations that are random with respect to adaptation, the resulting patterns may be random in the sense that there is no correlation between stylistic features and environmental variables affecting fitness. Only cultural variants in the bottom left extreme of Figure 1 (much of Sackett's isochrestic variation) will have such simple dynamics, but this case is of considerable interest here because the traditional style–function dichotomy holds that such dynamics should produce "random" patterns that are distinctively different from those characterizing variants at the bottom right of Figure 1. Even this simple comparison contains enough complexities to support the argument that archaeologists must pay closer attention to the details of process.

Imagine a very large, well-mixed population with a stylistic repertoire of n discrete elements, a transmission rule in which each individual acquires one of these variants at random, and a rule for innovation in which individuals (with some probability) switch to another variant with equal probability. A population using such rule will more or less rapidly converge to a state in which each variant is present in the population with equal frequency ($1/n$), no matter what the starting point.

Evolutionary systems with properties formally very similar to this kind of stylistic cultural variation have been extensively studied by population geneticists interested in what is called the *neutralism controversy*. The debate is briefly reviewed here because it contains important lessons for those interested in the evolution of stylistic cultural variation (for reviews of this subject see Kimura 1983; Gillespie 1987, 1991; Ridley 1993; see also Cavalli-Sforza and Feldman 1981 for theoretical applications to the special case of cultural variation). In the 1960s advances in molecular genetics demonstrated the existence of a huge amount of genetic variation in populations (Selander 1976). Individuals are heterozygous at *ca.* 5–15% of loci and at the population level 15–60% of loci are detectably polymorphic (at least one rare allele with a frequency greater than 1%). Kimura (1968) argued that natural selection could not possibly maintain so many polymorphic loci because recombination would ensure that each individual had a suboptimal genotype at many loci. Even a small amount of selection against each suboptimal locus would cumulatively ensure a huge selective load on the population. Kimura argued that such a large amount of variation could be maintained only if most alleles were neutral with respect to natural selection. The ensuing debate is of interest since isochrestic cultural variants are so similar in concept to neutral alleles in genetics. The problem in both cases is to distinguish traits under natural selection from traits that are not.

Tests pitting Kimura's neutralist claim against the alternative that selection plays a role in maintaining variation proceeded by the construction of models to deduce the unique predictions of the neutralist and selectionist hypotheses. The first complication here is that even in the completely neutral (i.e., "stylistic") case, one must take into account that populations are not infinite. Patterns in time and space will arise in finite populations if one includes the effects of genetic drift (random effects at the population level). New genes will be introduced into the population by mutation (random effects at the individual level), and the chances of "sampling" during reproduction in finite populations will cause some genes to increase and others to decrease by chance (random effects at the population level—genetic drift). For the mutation rates and population sizes thought to characterize animal populations, the theory predicts that many genetic loci should be monomorphic, but a fairly large proportion should have varying degrees of polymorphism. The data fit this prediction approximately, although there is considerable debate regarding the parameter values that must be assumed for mutation rates and population sizes. For example, for Drosophila, population sizes have to be rather small to account for the low levels of variation observed. Ohta (1976) argued that the fit is better if one assumes most alleles are subject to slight negative selection. Gillespie (1987, 1991) concurs that the Ohta version is the most empirically reasonable version of the neutral theory (albeit also flawed). At least aspects of the neutral theory can be rescued with other assumptions, for example, that populations were on average smaller in the Pleistocene. Regardless of the situation for genes, the theory may be quite appropriate for some kinds of cultural traits, especially isochrestic variants near the lower left of Figure 1.

If we survey a population over time, the neutral hypothesis predicts that there will be a more or less rapid turnover of genes as drift "selects" for at first one and then another genetic variant by chance. Superficially similar replacements will occur, however, if the locus concerned is responding to selection caused by environmental fluctuations or any other time-structured environmental factor. There is simply no warrant at this level for the archaeological assumption that random and selective effects will have qualitatively different patterns in time. Given error in sampling or random variation in the direction of selection, or some drift superimposed on a trajectory of selection, the gross time trend of neutral and selective evolution can be very similar. The processes of selection and random evolution by mutation and drift are sufficiently complex that models of both contain enough "tunable" parameters to mimic each other closely. This is one reason why a seemingly trivial debate could vex population genetics for a generation.

For the analogous case of human stylistic variants, the whole debate over parameter values would have to be conducted anew, but perhaps some guesses will give an idea of how random evolution proceeds. Let us start with a population in which one variant of a stylistic trait is overwhelmingly common. Individuals acquire their variant by copying someone of the parental generation at random, but also, rather rarely, certain individuals at random innovate one of the many other stylistic variants that are possible. Suppose some individual in a population of N individuals invents a particular new stylistic variant. Assuming, for simplicity, that each individual uses only one variant, what is the chance p that the new variant will eventually become in turn the overwhelmingly dominant variant in the population? It is simply,

$$p = \frac{1}{N} \tag{1}$$

which is easiest to see if we notice that by chance drift (random variation in the role of specific individuals in transmission each generation) in the long run, some one of the current stylistic alternatives will become the only one used (supposing no more innovation). Since every existing person's style has an equal chance of being the one that "drifts to fixation" as the population geneticist says, any given new innovation has a chance of being that lucky variant equal to its frequency at the point it first appears, which is (1) in the absence of simultaneous innovation. Of course, on average it will take a fair length of time for some given variant to be replaced by another by chance, and it is much more likely that any given innovation will be lost due to chance nonimitation

STYLE, FUNCTION, AND CULTURAL EVOLUTIONARY PROCESSES

of its originator or successors ($1-p$). If we suppose that there is a certain per-individual rate of innovation, u, we can ask what the rate of stylistic turnover in the population might be. In large populations there will be more innovations each generation, Nu, but according to (1) the rate at which they will become fixed is an inverse function of N. In this simple model the two exactly cancel, so that the time for one stylistic variant to replace another, k, is expected to be

$$\frac{1}{k} = Nu\left(\frac{1}{N}\right) = u \qquad (2)$$

That is, the turnover rate is just the reciprocal of the innovation rate.

If we imagine that societies are fairly conservative as regards stylistic innovation, say an innovation rate of a few tenths of a percent to a few percent per individual per generation, then the time to replace one style with another is a few hundred to a few tens of generations, independent of population size. That is, in a population in which a few tenths to a few percent of people innovate each generation, the turnover of stylistic features will occur on an archaeologically interesting time scale. A stylistic feature will drift in and out of a population over the course of hundreds or thousands of years, just like the standard battleship curves of stylistic seriation, as Neiman (1993) illustrates.

The spatial patterns generated by random stylistic choice will be governed by subpopulation size, innovation rates, and diffusion rates (see Neiman 1993). If population sizes are large and migration is high relative to innovation rates, chance effects alone will not be sufficient to cause populations to diverge. Of relevance here is the controversy in evolutionary biology as to whether genetic drift might be responsible for population differentiation and, in combination with group selection, play a role in moving populations across suboptimal troughs in the adaptive landscape, a famous hypothesis of Sewall Wright. The conditions are fairly restrictive in the biological case, but then mutation rates of genes are assumed to be very small, on the order of 10^{-6} per locus per generation.

If the corresponding innovation rates are something like 10^{-2} in the case of culture, and stylistic diffusion rates are not too high, chance stylistic differentiation of local populations is easy to imagine. In general, if innovation rates are greater than diffusion rates, we would expect chance differentiation to be important where selection is negligible (selection complicates the situation by retarding divergence between populations in which the same variants are favored, enhancing divergence between populations in which different variants are favored). The rate of differentiation will also depend on population size (Neiman 1993). Raw population size, however, will be less important than the portion of the population that is active in transmission of genes or culture, the "effective" population size in the jargon of evolutionary biology.

Cultural transmission is likely to be sensitive to effective population size because it often takes the form of "one-to-many" transmission, in which some traits are transmitted by relatively few "teachers" to large numbers of others. In this case, the effective population size is much smaller than a simple head count would indicate, which, *ceteris paribus*, strengthens the effect of drift (Cavalli-Sforza and Feldman 1981).

Barth (1987) gives the example from New Guinea of the Mountain Ok, among whom the transmission of ritual knowledge is controlled by the handful of older males in each community who have succeeded in passing through a long series of ritual initiations. Because the transmission of this ritual knowledge is infrequent and subject to errors of memory, innovation rates are much higher than they would be in genetic transmission. As a consequence, the esoteric lore of the semi-isolated Ok ritual communities diverges very rapidly. Ok shaman individually attempt to remain faithful to tradition, and, when they occasionally visit initiation rites in other communities, are shocked by the alarmingly large deviations from what they take to be ancient, immutable Ok truths.

Scientific disciplines are a more familiar case. Most modern disciplines count their practitioners and teachers in the thousands but are sharply stratified with respect to influence so that textbook writers, successful innovators, and individuals with many students have disproportionate weight. In the relatively narrow subdisciplines where most change is generated, the "effective" number of influential investigators in any one generation can be very small, perhaps less than ten, so chance effects in the evolution of science are perhaps likewise alarmingly likely. (We are indebted to J.R. Griesemer for this last example.)

Returning to the style–function dichotomy, the trouble is that there are no simple qualitative rules to distinguish these drift-induced patterns from those produced by simple adaptive processes like selection, by other adaptation-producing and nonadaptive cultural processes (outlined below), or by the interaction of several of these processes. To take a simple example, favorable technical innovations tend to occur at irregular intervals, and each sweeps through the population once discovered. The history of improvement of a technology thus tends to be characterized by a succession of improved forms in time. For this reason, particular technical forms can often be expected to conform to battleship curves that are indistinguishable from those that provide the basis for stylistic seriation (e.g., Phillips *et al.* 1951: Figure 11.3). We defy the reader to distinguish with respect to pattern Mangelsdorf's seriation of changing corn frequencies in the Tehuacan Valley, Mexico (Mangelsdorf 1974:Figure 15.23), which is presumably directed by selection, from Deetz and Dethlefsen's (1967:Figure 1) seriation of changing New England gravestone designs, which is presumably not directed by an adaptive process at all. In detail, these processes make quite different predictions about behavior. For example, a given technical improvement can sweep rapidly through even very large populations, whereas fixation by drift is a slow process in large populations. However, until we make reasonable estimates of the main parameters of the processes, such as innovation rates, magnitudes of selective differences, and effective sizes of populations, we cannot take advantage of the knowledge that selection and drift will produce different effects.

Pattern Resulting from Ordinary Adaptive Forces

The argument that stylistic variants must all have equal fitness rests on the implicit assumption that all adaptive problems have unique solutions. Stylistic variation must be neutral, the argument runs, because only neutral variation can persist. If two stylistic variants differed significantly in function, selection would rapidly eliminate the inferior variant. Persistent differences between groups in functional traits must then be the result of an environmental difference. This reasoning fails, however, if there are two or more locally stable traits. Natural selection is only a myopic optimizer—it causes a population to climb up the adaptive topography, eventually coming to rest at a local optimum. Most models of adaptive processes in cultural evolution suggest that they are similarly myopic (Cavalli-Sforza and Feldman 1981; Nelson and Winter 1982; Boyd and Richerson 1985; Durham 1991). If there is more than one local optimum, populations that begin from different positions may reach and maintain different equilibria even when some equilibria are better solutions than others. Clearly, environment alone cannot account for such differences.

There is substantial evidence that adaptive problems typically have many local optima. Engineers have shown that many design problems have this property. For examples, Kirkpatrick et al. (1983) report that where the problem was to minimize the number of slow connections between chips in the IBM 370, there are about $10^{1,503}$ possible arrangements, many of which are locally optimal. Local trial-and-error search cannot improve these local optima even though they tend to have about 4 times as many connections as the best arrangement the engineers discovered. Among the many local optima are a substantial number (≈ 70) of designs that are qualitatively different but essentially identical in function to the best arrangement found. Optimization texts (e.g., Wilde 1978) suggest that virtually all real world design problems "from dams to refrigerators" have many equilibria. We see no reason to suppose that the design problems facing people in subsistence economies are any different (Bettinger 1980).

Economists believe that increasing returns to scale, particularly those resulting from what they call "network externalities", often generate multiple evolutionary equilibria in modern economies (Arthur 1990). Network externalities arise when more widely available goods have an advantage merely because they are widely available. If you use a common make of computer, you have access to more software and more add-on hardware, more of your friends are able to help you learn to use it, and it is easier to collaborate with others. As a result, computer technologies with an initial numerical advantage may come to predominate even though they are inferior to alternative technologies. Similar phenomena likely occur in subsistence economies. It could be that the smaller !Kung arrow points really are better than the larger points made by the !Xo, but that !Xo who adopted smaller points would be worse off because they would be unfamiliar to exchange partners, harder to learn to make and use, and so on (see Weissner 1983).

Models suggest that many types of social interactions also lead to multiple evolutionary equilibria. The simplest examples are coordination games in which fitness is frequency-dependent but there is no conflict of interest among individuals (Sugden 1986). Driving on the left versus right side of the road is an example. It does not matter which side we use, but it is critical that we agree on one side or the other. Reciprocity provides a good example. Such models (Boyd and Richerson 1992b) suggest that there are a large number of different strategies that can capture at least some of the potential benefits of long-run cooperation. In order to persist when common, reciprocating strategies must retaliate against individuals who do not cooperate when cooperation is appropriate. When such a strategy is rare, it will interact mostly with other strategies which cooperate and expect cooperation in a different set of circumstances. Inevitably, a rare strategy will retaliate or suffer retaliation and cooperation will collapse. Thus, a common reciprocating strategy has an advantage relative to rare reciprocating strategies, even if the rare strategy would lead to greater long-run benefit were it to become common. Interactions of this kind are omnipresent in social life. Different social systems may often lead to variation in artifacts available in the archaeological record (e.g., Bettinger and Baumhoff 1982 and below). Systems with conical clan political organization will tend to have a minority of graves with rich furnishings whereas systems with segmentary lineages will tend toward a more egalitarian distribution of grave goods.

Pattern Resulting from Novel Adaptive Forces of Cultural Evolution

Culturally transmitted determinants of behavior are potentially subject to a number of evolutionary processes that Campbell (1965) terms "vicarious forces." These result from natural selection acting in the long run to produce decision rules that in turn vicariously select cultural variants. That is, individual choices about what traits to adopt and innovate will guide cultural evolution rather than selection acting directly on cultural variation, although the direct effects of selection are not necessarily negligible. There is not space here to give even a cursory review of the complexities that these forces engender (see Boyd and Richerson 1985 and Durham 1991). In principle, however, these forces have effects that are distinctively different from each other, from direct selection, and from those in systems affected only by random variation and drift.

Consider as an example the force that Boyd and Richerson call "conformist transmission." This is a version of frequency-dependent biased cultural transmission (Lumsden and Wilson 1980; Boyd and Richerson 1985: Chapter 7). "When in Rome, do as the Romans do" is a familiar example of a conformist or "positive" frequency-dependent rule. Conformist transmission causes people to discriminate against rare types, and is a potent suppressor of variation within societies. This can be quite adaptive in a spatially heterogeneous environment because it causes people to discriminate against migrants, who, more than locals, are prone to carry traits better adapted to other environments. On the other hand, it thwarts introduction of new variation, and so may impede adaptive tracking of environmental change over time. Such a simple decision rule may be most adaptive when it is applied without much judgment, as a kind of rule of thumb, to save on decision-making costs. Accordingly, conformist rules could be applied to wholly neutral traits as a by-product of their advantages with regard to adaptive traits, causing neutral and adaptive traits to pattern similarly.

The complexities introduced by such processes can be glimpsed in attempts to explain the observed spatial distribution of house forms in Africa. There, the ground plan of houses (rectangular, round, elliptical, and so forth) is highly variable from place to place but relatively uniform within individual societies. There is considerable spatial autocorrelation so that societies with similar house form tend to co-occur geographically. Cavalli-Sforza and Feldman (1981:209ff) argue that this pattern is perfectly consistent with driftlike effects in which (1) the low variation within societies is due to the one- to-many drift-enhancement effect, (2) the spatial autocorrelation is due to migration between closely adjacent groups, and (3) the differentiation of distant societies is due to isolation.

It is easy, however, to produce a counterscenario that couples a different form of transmission with adaptation/selection. Imagine that house builders use a "biased sampling" rule for acquiring cultural traits in which they survey a number of cultural models and imitate the house form most common among those models. As Africans adopted agriculture and began to build houses a few thousand years ago, subtly different house types may have been advantageous in different places, perhaps because of differences in raw materials from place to place. Such early accidents could have been frozen by conformist transmission, and distributed about the landscape by migration or by the tendency of non-house builders to acquire houses from nearest neighbors. Alternatively, adaptive considerations such as microclimate or availability of building materials may have tended to determine the standard house form in particular locations and thereafter conformity acted to suppress variation around that standard. Both hypotheses differ from that of Cavalli-Sforza and Feldman in that house form is determined initially by adaptation and subsequently by conformist transmission, i.e., by a mixture of adaptation and transmission.

Nonconformist transmission rules and more complex forms of "trend-watching" are quite conceivable (Lumsden and Wilson 1980). The nonconformist version of frequency-dependent transmission will protect variants from loss within societies by drift, tending to preserve variation arising by individual invention. This process will mimic a situation with high innovation and migration rates plus drift. Again and again the point emerges that empirical patterns, even the most rigidly structured ones, are often consistent with a variety of different processual hypotheses. It is unrealistic to expect to be able at a glance to segregate them unambiguously as resulting from either selection or neutral transmission.

Patterning as the Result of Correlations among Characters

One possible way to distinguish adaptive from stylistic-neutral patterns of variation is by the presence of plausible selective factors capable of explaining the observed pattern. In the genetic case, patterns related to selective factors (e.g., climate) are often found (e.g., Clegg and Allard 1972; Watt 1977). However, as proponents of the neutral theory countered, linkage of adaptive and neutral genes could easily give rise to patterns of neutral alleles that are indistinguishable from those of adaptive variants. A neutral allele at one locus can "hitchhike" to high frequency if it is statistically associated (linked) to an adaptive variant at another locus. In the case of genes, the statistical association is generally assumed to result from physical proximity on the chromosome, so that if a gene for hair color is located on the same chromosome and very near the gene for cold tolerance, the pattern of hair color (neutral) might end up being closely associated with patterns of climate due to selection on a linked gene influencing limb length or some other direct adaptation to cold.

Important technical innovations that produce waves of population expansion could easily drag a host of neutral or near-neutral genetic and cultural traits to high frequency because of a chance high frequency in the population which first acquires the adaptive trait. Physical linkage analogous to genetic linkage may be involved. For example, a complex tool with many parts may be learned more or less as a whole, so that its individual components will seldom "recombine." An adaptive innovation in one part of a tool may cause the hitchhiking of nonadaptive variation and stylistic features with regard to other parts.

Hitchhiking, however, does not necessarily require any linkage in the physical sense, only an initial statistical association. Because of this, genes can easily hitchhike with cultural innovations or vice versa. Thus, Ammerman and Cavalli-Sforza (1985) explain the gradient of certain human gene frequencies in Europe as a result of the genes hitchhiking on the wavelike spread of agriculture from the Middle East west-northwestwards beginning about 9000 B.P. It is difficult to distinguish their hypothesis from the selective explanation that these gradients are largely adaptations to climatic gradients, since measures of climate (e.g., isotherms) largely parallel the isolines for the dates of the agricultural wave. The hitchhiking hypothesis has been frequently invoked in various forms to explain the spread of languages, especially Indo-European (Renfrew 1987; Mallory 1989). Renfrew's hypothesis is that Indo-European hitchhiked from an original focus in Anatolia, like Ammerman and Cavalli-Sforza's genes, with the spread of agriculture. Mallory discusses the more traditional hypotheses that link Indo-European to later improvements in the use of horses in warfare. Given that language variation is prototypically stylistic, with no functional difference between alternative words, etc., the patterning of language in time and space is a powerful confirmation of the importance of the hitchhiking effect. Such cultural hitch-hiking, of course, is the source of what is known as "Galton's problem," in which correlation produced by adaptive forces cannot be distinguished from correlations produced by shared history. Deetz and Dethlefsen (1967) have archaeologically documented a form of hitchhiking in New England gravestone styles that is evidently related to shifting trade networks.

Patterning as the Result of Signaling

As we noted in reference to Weissner's (1983, 1985) work, anthropologists and archaeologists commonly attribute communicative functions to stylistic variation, and at least the more symbolic cases of style in artifacts do commonly appear to function as expressive or emblematic communication. What is less well appreciated is that the processes that affect the evolution of communicative elements of style go well beyond simple random innovation and statistical drift. Rather, several different directional evolutionary forces will affect stylistic variables.

The issue is not a simple one. Consider the prototypical symbolic communication system, human language. All human languages are functionally equivalent (variations in technical vocabulary aside). It does not matter which one we speak, but it is important for purposes of efficient communication that we follow local conventions of semantics and syntax. Thus, to preserve function, we might expect forces that act to limit individual-level innovation (e.g., the conformist transmission bias) to dominate the evolution of language. If so, we would expect language evolution to be quite conservative when, in fact, it is fairly rapid. Mutually unintelligible dialects arise in separate populations in a few hundred years (Ruhlen 1994). At first glance, this rapid Tower of Babel evolution seems to be in defiance of the communication function of language. Why don't human communication systems behave in a much more conservative fashion? For that matter, why isn't language a hard-wired human universal? The highly conserved basic structure of our genetic code behaves as expected, but our language does not.

The most plausible current hypothesis to explain the rapid evolution of human symbolic systems is that their main function is to communicate emblematic information about group membership and about appropriate group behavior in cases where individuals are frequently exposed to social interaction with members of another group.

Detailed microevolutionary studies of dialect change by sociolinguists (Labov 1980) support this idea. In many areas of the contemporary world, microdialect change is rapid enough to be detected between generations. Dialect change seems to be set in motion by sociological processes, for example, competition between ethnic groups. In one of Labov's cases, the White dialect of Philadelphia appears to have arisen in response to the influx of Southern Blacks during and after WWII. In another case, on Martha's Vineyard, the evolution of Islander dialect appears to be driven by Islander desire to emphasize an identity separate from mainland tourists toward whom economic necessity compels an uncomfortable level of deference. Cohen (1974) developed a very similar hypothesis to explain the evolution of ideological and ceremonial systems, such as the adoption of Freemasonry by Sierra Leone Creoles in the face of political competition from traditionally disenfranchised groups.

By preserving ethnic identity, this sort of process does foster social solidarity but need not be viewed in purely structural-functional terms because the symbolic behaviors that identify group membership may often be associated behaviors that are functionally adaptive. Boyd and Richerson (1985 Chapter 8, 1987; Richerson and Boyd 1989) have examined this possibility with models of the evolution of symbolic cultural traits inspired by data such as Labov's and Cohen's. In the simplest systems they have studied, populations are characterized by a symbolic trait, such as dialect, which is selectively neutral, and an ordinary adaptive character, such as a subsistence technique. Both traits were modeled as quantitative characters. They suppose that children acquire the symbolic variant when young by unbiased imitation of a local adult. In a second episode of imitation as "teenagers," individuals acquire their subsistence trait by observing and imitating a wider range of individuals. They bias this second decision about whom to imitate in favor of individuals bearing a symbolic variant similar to theirs. After a period of experimentation, these "teenagers" compare the success of different behavioral combinations (symbolic plus adaptive) and reject less successful combinations in favor of more successful ones. The criterion of "success" is arbitrary in the model, but can certainly be interpreted as adaptive success.

According to this model, in a spatially variable world in which optimal subsistence behavior is very different in different environments, a correlation can build up between the subsistence trait and a symbolic marker, so long as the rate of migration of people in the first symbolic episode of cultural transmission is less than in the second episode where subsistence traits are also transmitted. Once a correlation accumulates between a symbolic trait and a favored subsistence trait, there is a substantial advantage to using the symbolic marker as a guide for whom to imitate. Simulations show that two populations using the symbolic rule will diverge with regard to the indicator character until the mean values of the adaptive character are optimal, whereas the adaptation to a variable environment by nonsymbolic populations is adversely affected by migration and leads to less successful adaptations. Boyd and Richerson argue that such models are consistent with the hypothesis of a widespread advantage for the use of affect-laden emblematic symbol systems to regulate cultural transmission.

What Weissner calls assertive style may result from the buildup of correlations between stylistic and functional variables. Several careful studies of contemporary populations (Irons 1979; Borgerhoff Mulder 1987) have shown a strong correlation between prestige, as defined by the local ideological system, and wealth and reproductive success. This correlation also suggests that the use of symbolically defined status as a guide to whom to imitate would be functional, as Flinn and Alexander (1982) argued. Boyd and Richerson (1985 Chapter 8) review several other lines of evidence for the important role of using marker traits in choosing from whom to acquire cultural variants. Empirical microevolutionary studies of expressive art styles in the modern West have been conducted by Martindale (1975, 1990). He gives an interesting account of how psychological processes might drive the trends and cycles he discovers in his data.

The processes that build correlations between arbitrary stylistic features and adaptive characters might be termed "active hitchhiking." Again, it would not be easy to distinguish the patterns generated by this process from simpler ones, especially from the effects of "passive" hitchhiking. The potential for confounding is even more serious here since those aspects of the symbol system that are most subject to drift tend to make the best adaptive markers. This is because the active hitchhiking effect is weak when the symbolic difference between populations is small. The biased imitation effect works on the correlation between marker and adaptive characters (Boyd and Richerson 1987). Accordingly, if the initial variation between populations is small, the correlation between symbolic and adaptive characters built up by migration between them will also necessarily be small. When there is a substantial correlation between symbolic marker and adaptive character in a given environment, individuals who have the common symbolic value associated with an advantageous variant of the adaptive character are doubly advantaged in cultural transmission relative to other types. Contrariwise, before some variation in the symbolic character arises, the tendency to build further correlation will be very weak. For example, if an ancestral population divides and becomes segregated in different habitats, the bias process can build up a symbolic difference between them only very slowly. If random, driftlike processes are strong in some characters, they will provide the first perceptible differences between the descendant populations. The active hitchhiking effect is then liable to seize just these traits and build a correlation between them and adaptive characters. It is under such conditions, for example, that chance local variation in the frequency of functionally equivalent technical alternatives might become the basis for symbolic variation between groups.

The assertive use of style by individuals, particularly by prestigious individuals, seems to be the cause of rapid evolution of potential marker traits. Labov's account of dialect evolution and Martindale's account of artistic evolution both depend on a certain limited taste for the novel which drives linguistic and artistic evolution in spite of the forces of conformity that are required to keep such systems functioning as media of communication. Language cannot change much in any one generation without disrupting communication between individuals, but the small innovations made by the leaders of linguistic change lead to rather dramatic changes in a relatively few generations. We hypothesize that the assertive use of style is the motor that builds up variation between semi-isolated groups, which can in turn then serve as badges of group membership. The evolutionary origins of the modern human sense of style might well lie in the advantage of using stylistic variation as an indicator of differences in adaptive characters.

Patterning as a Result of Runaway Effects

The process of using a marker trait to choose whom to imitate has potentially explosive unstable dynamic properties (Boyd and Richerson 1985 Chapter 8; Richerson and Boyd 1989). In the abstract it is easy to see that a dramatically exaggerated prestige system can be protected against natural selection or selection-derived rational choice if success in the prestige system offsets losses related to the maladaptive consequences of the exaggeration. Once enough people use a specific marker to choose whom to imitate, anyone failing to display that indicator will be effectively ignored in the process of cultural transmission. Where social systems are based on an element of coercion, the possibility of maintaining arbitrary, non-functional behaviors is increased still further (Boyd and Richerson 1992b). The key question is how such a set of preference rules favoring imitation of people displaying costly prestige symbols, or a willingness to punish deviants, can arise in the first place. As with the analogous case of mate-choice sexual selection, theoretical models show that a system of coupled preference characters and display characters can run away to exaggerated extremes (Lande 1981). The male-biased display characters of many animals, such as the feathers of peacocks and the elaborate constructions of bowerbirds, are often attributed to the runaway effect. We regard the similarity between the plumes of birds and the finery displayed by prestigious humans as more than coincidental. In this respect, the runaway hypothesis provides a way of turning Sahlins's (1976) notion of "cultural reason" into a cogent formal argument.

The importance and mechanics of the runaway effect are hotly debated by evolutionary biologists (see, e.g., Kirkpatrick *et al.* 1990; Barton and Turelli 1991). Barton and Turelli's theoretical investigations suggest that the pure runaway effect is weak because the forces maintaining the correlation between the symbolic display and preference traits are inherently weak. What will encourage the exaggeration process is some independent adaptive advantage accruing to the selection of individuals with elaborate markers. For example, if economic success generates the wherewithal to display status more effectively (buy the fanciest car, pay the bride-price for the youngest and most beautiful woman), there will remain a correlation between ordinary adaptive success and the degree of exaggeration of display. Then it pays in both the ordinary adaptive and prestige games to choose mates or mentors with the most exaggerated system of prestige. In the limit, all of the gains accruing from ordinary adaptive advantages are dissipated in support of the most elaborate possible status displays, a sort of perverse inversion of the ordinary hitchhiking effect. In such cases, it may be said justifiably that the culturally driven symbolic system has captured the mundane economic system, much as Sahlins claims.

Clearly, the gross patterns predicted by the runaway and signaling hypotheses are rather similar. Boyd and Richerson (1985 Chapter 8) argue that the ordinary adaptive advantages of choosing mentors by means of indicator characters will maintain this sort of choice mechanism by natural selection, even though the system misfires occasionally and gets caught up in the runaway process. They further suppose that the traits most subject to exaggeration will generally be ones historically connected with adaptation. The growing of giant yams on Ponape as a part of prestige contests is a possible example. It seems plausible that when the custom originated, good farmers did grow larger yams, and that large yams were a good index of yam-growing talent. If Barton's and Turelli's hypothesis applies in the cultural case, the growers of giant yams may still be the best horticulturalists; the adaptive and runaway hypotheses are really wonderfully entangled in this case.

CONCLUSION

The main attraction of the style–function dichotomy is that it apparently reduces the task of archaeological explanation to a manageable subset of phenomena that can be addressed by simple causal models. Some will undoubtedly read our rejection of the dichotomy as complicating the problem of archaeological interpretation to the point of impossibility. Because we reject the dichotomy and even cite the works of such authors as Sahlins with (qualified) approval, still others will read us as advocating a form of poststructuralist archaeology. Neither reading is warranted. We are enthusiastic advocates of causal and materialist explanations of social phenomena. We are also advocates of simple models of complex phenomena (Levins 1966). Treating patterned variation in artifacts and behaviors as though it were purely functional and adaptive and assuming that stylistic variation is noisy and irrelevant may often be an acceptable simplification. Surely, human groups cannot exist as going economic concerns unless a large fraction of patterned variation is adaptive. Likewise, assuming that style behaves as if it were subject only to random innovation and drift provides an important theoretical warrant for seriation that can be highly useful even when patterns depart somewhat from the ideal. Within this context, our argument boils down to these three simple points.

First, there is good reason to think that the style–function dichotomy is frequently an unacceptable simplification. It is well worth thinking about this possibility and what it implies about how we should do archaeology. As Wimsatt (1980) notes in his defense of the use of simple models, failure to recognize the specific limitations of widely adopted simplifying assumptions can lead to dangerous overconfidence in the robustness of our models and their results. As part of this, we must squarely face the difficulties involved in solving what physical scientists call the inverse problem (more familiar to archaeologists as the problem of "equifinality"). It may be hard, sometimes perhaps impossible, to infer the microscale processes that gave rise to a particular macroscale pattern. Many different evolutionary processes, for instance, can cause the familiar battleship (lenticular) pattern of increase and decrease. The problems this raises cannot be ignored. It is difficult to distinguish isochrestic from functional variation, as nearly everyone agrees. However, if the argument we have presented is correct, assuming that function and style can be separated into discrete categories with very different evolutionary properties is not possible. The processes of evolution are just more complex than that. What is called for is a methodologically rigorous program of study patterned in the mold of contemporary middle range studies that use tightly controlled ethnoarchaeological and taphonomic investigation to distinguish the signatures of different processes that tend to produce outwardly similar archaeological consequences. In some cases the data will be insufficient to decide between competing alternative hypotheses but there is no reason to think that this will be true generally. The inverse problem is cause for despair only if it can be solved so infrequently that there is no hope of building a satisfactory picture of the relative power of different general hypotheses. Otherwise, it is merely a challenge to our imagination and initiative.

Second, style is too important and too interesting to leave to structuralists and postmodernists. A number of important archaeological phenomena make much more sense if we assume that stylistic variation is functionally important. Why was the Upper Paleolithic transition a stylistic as well as economic revolution? Why does state formation so frequently involve the elaboration of religious institutions, ideology, and the arts? How costly are symbolic institutions, and how much do they distort or foster adaptation (however that might be operationalized)?

Third, Darwinian theory will eventually offer a processual account of cultural evolution that is as powerful as the one it now offers for genetic evolution. The problem presently is that we have limited knowledge of the operation of these processes in the cultural realm and are handicapped in our ability to use them in interpreting past behavior. On the other hand, this situation should be attractive to those of us who continue to share the processual goals that inspired the New Archaeology. There are a large number of essentially unstudied processes begging the kind of critical experimental and observation program advocated by Binford and others.

Archaeologists, who are sometimes driven to do the work that should more properly fall to ethnographers (ethnoarchaeology), should appreciate that archaeology must play a distinctively critical role in understanding the processes of cultural evolution. The synchronic study of symbol systems and their evolution on the micro time scale is surely critical, but archaeologists, historians, and paleoanthropologists have a monopoly on data from the longer time scales over which the evolutionary processes generally work themselves out. The models of the adaptive role of symbolic marking of ethnic groups reviewed above, for example, are necessarily silent about just what sorts of adaptive differences between groups might be protected by this mechanism. That depends on how easily correlations between various kinds of traits can be built up in the face of migration. It is unlikely that short-term studies will be as convincing in this regard as the actual long-run data.

A bit of our own work illustrates the kind of process-related information available from long records. Bettinger and Baumhoff (1982) have examined the case of the spread of Numic speakers across the Great Basin of the western United States from about 700 to 200 years ago. In this case the evidence supports the idea that the ethnic boundary between Numic peoples and their pre-Numic predecessors must have limited the spread of a social-organizational variable, not a direct technological variable. The record indicates that the same technology and tool types were everywhere available, yet the spread of stylistic elements associated with Numic speakers, including language and ideology, is associated with higher densities, differences in location of settlements, and quantitative variation in frequencies of the various tool types. Since the more plant-intensive Numic strategy required a larger role for plant storage and women's labor, it is plausible that the key to Numic success was a normative complex that condoned the hoarding of plant resources, gave women a greater role in decision-making, and reduced the autonomy of hunters.

The utilitarian consequences of mundane technology (e.g., seed-beaters) are generally rather obvious, and hence move easily across boundaries. Social norms have more complex and far-reaching effects that are often difficult for actors to understand and more closely tied to affect-laden ideological systems. Perhaps only an ethnic isolate is likely to take unusual steps away from obligate sharing and toward gender egalitarianism. Once the ethnic advance begins, an ethnic boundary can explain why the losing group persists in retaining its behavior despite the obvious disadvantages.

The Numic case may or may not be correctly interpreted, and even if correctly interpreted, it may not be representative. It does have the virtue of suggesting testable hypotheses: (1) that ethnic or other style-marked boundaries are important in the origin and spread of certain types of innovations and (2) that technical innovations *per se* are likely to spread irrespective of style-marked boundaries, whereas more subtle aspects of adaptation (as judged by ability to support higher population densities for example) are likely to require boundaries to originate, and are then likely to spread associated styles by active hitchhiking. It is archaeologists who are in a position to probe the long-term patterns of correlation between different kinds of traits, and hence to make an essential contribution to the very basic social science problem of understanding just what is the significance of modern humans' massive preoccupation with style, and how it is that we came to replace populations with an (apparently) more narrowly utilitarian outlook.

REFERENCES

Ammerman, A.J., and Cavalli-Sforza, L.L., 1985, *The Neolithic Transition and the Genetics of Populations in Europe*, Princeton University Press, Princeton.

Arthur, W.B., 1990, Positive Feedbacks in the Economy, *Scientific American* February:92–99.
Barth, F., 1987, *Cosmologies in the Making: A Generative Approach to Cultural Variation in Inner New Guinea*, Cambridge University Press, Cambridge.
Barton, N.H., and Turelli, M., 1991, Natural and Sexual Selection on Many Loci, *Genetics* 127:229–255.
Bettinger, R.L., 1980, Explanatory-Predictive Models of Hunter-Gatherer Behavior, in: *Advances in Archaeological Theory and Method*, Vol. 3 (M.B. Schiffer, ed.), Academic Press, New York, pp. 189–255.
Bettinger, R.L., 1991, *Hunter-Gatherers: Archaeological and Evolutionary Theory*, Plenum Press, New York.
Bettinger, R.L., and Baumhoff, M.A., 1982, The Numic Spread: Great Basin Cultures in Competition, *American Antiquity* 47:485–503.
Binford, L.R., 1962, Archaeology as Anthropology. *American Antiquity* 48:217–225.
Binford, L.R., 1965, Archaeological Systematics and the Study of Culture Process, *American Antiquity* 31:203–210.
Binford, L.R., 1983, *In Pursuit of the Past*, Thames and Hudson, London.
Binford, L.R., 1989, Styles of Style, *Journal of Anthropological Archaeology* 8:51–67.
Borgerhoff Mulder, 1987, On Cultural and Biological Success: Kipsigis Evidence, *Anthropologist*, 89:619–634.
Boyd, R., and Richerson, P.J., 1985, *Culture and the Evolutionary Process*, University of Chicago Press, Chicago.
Boyd, R., and Richerson, P.J., 1987, The Evolution of Ethnic Markers, *Cultural Anthropology* 2:65–79.
Boyd, R., and Richerson, P.J., 1992a, How Microevolutionary Processes Give Rise to History, in: *History and Evolution* (M.H. and D.V. Nitecki, eds.), State University of New York, Albany, pp. 179–209.
Boyd, R., and Richerson, P.J., 1992b, Punishment Allows the Evolution of Cooperation (Or Anything Else) in Sizable Groups, *Ethology and Sociobiology* 13:171–195.
Campbell, D.T., 1965, Variation and Selective Retention in Sociocultural Evolution, in: *Social Change in Developing Areas: A Reinterpretation of Evolutionary Theory* (H. Barringer, G.I. Blanksten, and R.W. Mack, eds.), Schenkman, Cambridge, MA, pp. 19–49.
Cavalli-Sforza, L.L., and Feldman, M.W., 1981, *Cultural Transmission and Evolution: A Quantitative Approach*, Princeton University Press, Princeton.
Clegg, M.T., and Allard, R.W., 1972, Patterns of Genetic Differentiation in the Slender Wild Oat, Avena barbata, *Proceedings of the National Academy of Sciences of the United States of America* 69:1820–1824.
Cohen, A., 1974, *Two-Dimensional Man: An Essay on the Anthropology of Power and Symbolism in Complex Society*, University of California Press, Berkeley.
Cosmides, L., and Tooby, J., 1992, Cognitive Adaptations for Social Exchange, in: *The Adapted Mind: Evolutionary Psychology and the Generation of Culture* (J.H. Barkow, L. Cosmides, and J. Tooby, eds.), Oxford University Press, Oxford, pp. 163–228.
Dawkins, R., 1986, *The Blind Watchmaker*, Norton, New York.
Deetz, J.F., and Dethlefsen, E.S., 1967, Deaths Head, Cherub, Urn and Willow, *Natural History* 76:29–37.
Dunnell, R., 1978, Style and Function: A Fundamental Dichotomy, *American Antiquity* 43:192–202.
Dunnell, R., 1980, Evolutionary Theory and Archaeology, in: *Advances in Archaeological Theory and Method*, Vol. 3 (M.B. Schiffer, ed.), Academic Press, New York, pp. 35–99.
Durham, W.H., 1991, *Coevolution: Genes, Culture, and Human Diversity*, Stanford University Press, Stanford.

Eberhard, W.G., 1986, *Sexual Selection and Animal Genitalia*, Harvard University Press, Cambridge, MA.
Flinn, M.V., and Alexander, R.D., 1982, Culture Theory: The Developing Synthesis from Biology, *Human Ecology* 10:383–400.
Gillespie, J.H., 1987, Molecular Evolution and the Neutral Allele Theory, *Oxford Surveys in Evolutionary Biology* 4:10–37.
Gillespie, J.H., 1991, *The Causes of Molecular Evolution*, Oxford University Press, Oxford.
Hamilton, W.D., and Zuk, M., 1982, Heritable True Fitness and Bright Birds: A Role for Parasites, *Science* 218:384–387.
Harris, M., 1968, *Rise of Anthropological Theory*, Crowell, New York.
Heiner, R.A., 1983, The Origin of Predictable Behavior, *American Economic Review* 73:560–595.
Irons, W., 1979, Cultural and Biological Success, in: *Evolutionary Biology and Human Social Behavior: An Anthropological Perspective* (N. Chagnon and W. Irons, eds.), Duxbury, North Scituate, MA, pp. 257–272.
Kimura, M., 1968, Evolutionary Rate at the Molecular Level, *Nature* 217:624–626.
Kimura, M., 1983, *The Neutral Allele Theory of Molecular Evolution*, Cambridge University Press, Cambridge.
Kirch, P.V., 1980, The Archaeological Study of Adaptation: Theoretical and Methodological Issues, in: *Advances in Archaeological Theory and Method*, Vol. 3 (M.B. Schiffer, ed.), Academic Press, New York, pp. 101–156.
Kirkpatrick, M., 1989, Is bigger always better?, *Nature* 337:116–117.
Kirkpatrick, M., Price, T., and Arnold, S.J., 1990, The Darwin–Fisher Theory of Sexual Selection in Monogamous Birds, *Evolution* 44:180–193.
Kirkpatrick, S., Gelatt, C.D., and Vecchi, M.V., 1983, Optimization by Simulated Annealing, *Science* 220:671–680.
Kroeber, A.L., 1948. *Anthropology*, Harcourt, Brace & World, New York.
Kroeber, A.L., and Richardson, J., 1940, Three Centuries of Womens Dress Fashions: A Quantitative Analysis, *University of California Anthropological Records* 5(2).
Labov, W. (ed.), 1980, *Locating Language in Time and Space*, Academic Press, New York.
Lande, R., 1981, Models of Speciation by Sexual Selection on Polygenic Traits, *Proceedings of the National Academy of Sciences of the United States of America* 78:3721–3725.
Leonard, R.D., and Jones, G.T., 1987, Elements of an Inclusive Evolutionary Model for Archaeology, *Journal of Anthropological Archaeology* 6:199–219.
Levins, R., 1966, The Strategy of Model Building in Population Biology, *American Scientist*. 54:421–431.
Lumsden, C.J., and Wilson, E.O., 1980, Translation of Epigenetic Rules of Individual Behavior into Ethnographic Patterns, *Proceedings of the National Academy of Sciences of the United States of America* 77:4382–4386.
Mallory, J.P., 1989, *In Search of the Indo-Europeans: Language, Archaeology and Myth*, Thames and Hudson, London.
Mangelsdorf, P.C., 1974, *Corn*, Harvard University Press, Cambridge, MA.
Martindale, C., 1975, *Romantic Progression: The Psychology of Literary History*, Halsted Press, New York.
Martindale, C., 1990, *The Clockwork Muse: The Predictability of Artistic Change*, Basic Books, New York.
Maynard Smith, J., 1991, Theories of Sexual Selection, *Trends in Ecology and Evolution* 6:146–151.
Neiman, F., 1993, What Makes the Seriation Clock Tick? Paper presented at the Annual Meeting of the Society for American Archaeology, St. Louis.
Nelson, R.R., and Winter, S.G., 1982, *An Evolutionary Theory of Economic Change*, Harvard University Press, Cambridge, MA.

O'Brien, M.J., and Holland, T.D., 1992, The Role of Adaptation in Archaeological Explanation, *American Antiquity* 57:36–59.

O'Connell, J.F., Jones, K.T., and Simms, S., 1983, Some Thoughts on Prehistoric Archaeology in the Great Basin, in: *Man and Environment in the Great Basin* (D.B. Madsen and J.F. Connell, eds.), Society for American Archaeology Papers No. 2, Washington, DC, pp. 227–240.

Ohta, T., 1976, Role of Very Slightly Deleterious Mutations in Molecular Evolution and Polymorphism, *Theoretical Population Biology* 10:254–275.

Opler, M.E., 1964, The Human Being In Culture Theory, *American Anthropologist* 66:507–528.

Phillips, P., Ford, J.A., and Griffin, J.B., 1951, *Archaeological Survey in the Lower Mississippi Alluvial Valley, 1940–1947*, Papers of the Peabody Museum No. 25, Harvard University, Cambridge, MA.

Pomiankowski, A., 1988, The Evolution of Female Mate Preferences for Male Genetic Quality, *Oxford Surveys in Evolutionary Biology* 5:136–184.

Renfrew, C., 1987, *Archaeology and Language: The Puzzle of Indo-European Origins*, Cambridge University Press, Cambridge.

Richerson, P.J., and Boyd, R., 1989, The Role of Evolved Predispositions in Cultural Evolution: Or, Human Sociobiology Meets Pascal's Wager, *Ethology and Sociobiology* 10:195–219.

Ridley, M., 1993, *Evolution*, Blackwell, Oxford.

Ruhlen, M., 1994, *The Origin of Language: Tracing the Evolution of the Mother Tongue*, Wiley, New York.

Ryan, M.J., Fox, J.H., Wilczynski, W., and Rand, A.S., 1990, Sexual Selection in the Frog Physalaemus pustulosus, *Nature* 343:66–67.

Sackett, J.R., 1982, Approaches to Style in Lithic Archaeology, *Journal of Anthropological Archaeology* 1:59–112.

Sackett, J.R., 1985, Style and Ethnicity in the Kalahari: A Reply to Weissner, *American Antiquity* 50:154–159.

Sackett, J.R., 1986, Style, Function, and Assemblage Variability: A Reply to Binford, *American Antiquity* 51:628–634.

Sahlins, M., 1976, *Culture and Practical Reason*, University of Chicago Press, Chicago.

Selander, R.K., 1976, Genetic Variation in Natural Populations, in: *Molecular Evolution* (F.J. Ayala, ed.), Sinauer, Sunderland, pp. 21–45.

Simon, H.A., 1959, Theories of Decision-making in Economics, *American Economic Review* 49:253–283.

Stringer, C., and Gamble, C., 1993, *In Search of the Neanderthals*, Thames and Hudson, London.

Sugden, R., 1986, *The Economics of Rights, Co-operation and Welfare*, Blackwell, Oxford.

Steward, J.H., 1938, *Basin-Plateau Aboriginal Sociopolitical Groups*, Bureau of American Ethnology Bulletin 120, Washington, DC.

Vayda, A.P., and Rappaport, R.A., 1968, Ecology: Cultural and Non-Cultural, in: *Introduction to Cultural Anthropology* (J.A. Clifton, ed.), Houghton, Boston, pp. 477–497.

Watt, W.B., 1977, Adaptation at Specific Loci: I. Natural Selection on Phosphoglucose Isomerase of Colias Butterflies: Biochemical and Population Aspects, *Genetics* 87:177–194.

Weissner, P., 1983, Style and Social Information in Kalahari San Projectile Points, *American Antiquity* 48:235–276.

Weissner, P., 1985, Style or Isochrestic Variation? A Reply to Sackett, *American Antiquity* 50:160–166.

West-Eberhard, M.J., 1983, Sexual Selection, Social Competition, and Speciation, *Quarterly Review of Biology* 58:155–183.

White, L.A., 1959, *The Evolution of Culture*, Farrar, Straus, New York.

Wilde, D.J., 1978, *Globally Optimal Design*, McGraw–Hill, New York.

Wimsatt, W., 1980, Reductionistic Research Strategies and Their Biases in the Units of Selection Controversy, in: *Scientific Discovery, II. Case Studies* (T. Nickles, ed.), Reidel, Dordrecht, pp. 213–259.

Chapter 9

In Search of the Watchmaker
Attribution of Agency in Natural and Cultural Selection

PAUL GRAVES-BROWN

> *Natural selection, the blind, unconscious, automatic process which Darwin discovered, and which we now know is the explanation for the existence and apparently purposeful form of all life, has no purpose in mind. It has no mind and no mind's eye. It does not plan for the future. It has no vision, no foresight, no sight at all. If it can be said to play the role of the watchmaker in life, it is the blind watchmaker.* (Dawkins 1986:5)

INTRODUCTION

Recognition of the relationship between natural selection and cultural change has a longer history than most people think (Costall 1991). Charles Darwin and many of his contemporaries were well aware of the analogy between change in artifact form and natural processes of change in living things. Yet, despite this considerable history of study, recent attempts to apply Darwinian theory to human culture have failed to confront problems faced by Darwin himself.

In this paper I begin by exploring the idea of selection in its application to nature. In particular, I shall concentrate on the attribution of agency in evolution; to ask who or what is the "blind watchmaker" in the process of natural selection. I argue that the term *selection* has inappropriate overtones of metaphysical intervention and conclude that the term *struggle* is more apposite in allowing for the active role of organisms themselves.

PAUL GRAVES-BROWN • Department of Psychology, University of Southampton, Highfield, Southampton SO17 1BJ, United Kingdom.

Turning to accounts of cultural selection, it then becomes clear why the analogy with Darwinian theory has not always been successful. This too is a problem of the attribution of agency. In my view, overemphasis on the blind determinism of natural selection is complemented by an excessive reliance on free will or choice in models of cultural selection. Lacking any equivalent of the natural process of struggle, these models have no means of accounting for the differential "success" of cultural practices.

THE VERB "TO SELECT"

to select. (verb, transitive). choose carefully, choose as best or fittest.

In English at least, no verb can describe an actorless act.[1] The verb *to select*, being transitive, has both an object—that which is chosen carefully, as the best or fittest—and a subject—the chooser, the selector. Ostensibly, in both natural and cultural processes of selection, there is little doubt about the objects of selection. Organisms or ideas are selected as the "best or fittest" (and here perhaps the dictionary is itself influenced by Darwinism). But the selecting *subject*, the agent of selection, remains obscure and at times controversial.

To a considerable extent the root of this problem lies in the history of evolutionary theory. In the *Origin*, Darwin chose studies of the artificial selection of domesticates to lead into his discussion of natural selection.[2] Quite possibly he intended this device to ease his readership into the ideologically difficult concept of an evolutionary process without divine intervention (Young 1985), but the implications of the artificial selection analogy were to haunt him for the rest of his career.

The principal achievement of Darwinian theory is surely the elimination of the "watchmaker" in the explanation of the origins and development of life. Darwinism has scientific credibility because it appears to eliminate metaphysical causes from nature. Yet, despite its dismissal of divine intervention, Darwin's theory grew out of the metaphysical tradition of Natural Theology. Thus, as Young (1985:87) states; "In moving from artificial to natural, Darwin retains the anthropomorphic concept of selection, with all its voluntarist overtones."

The very choice of the verb *to select* carries with it an anthropomorphic sense of volition and action. As Beer (1983:68) remarks, "In the first edition of *The Origin* both Nature and natural selection have grammatically the function of agents—and, moreover, despite his later exasperation with the issue, Darwin does endow them with conscious activity...." To quote a classic example:

---

[1] As Erasmus Darwin originally observed, English has a strong tendency to personification of objects derived from its essentially ungendered structure (Beer 1983).
[2] And, in a sense, Mendel's theory of inheritance followed a similar path from his experiments with domesticated plants.

It may be said that natural selection is daily and hourly scrutinising, throughout the world, every variation, even the slightest; rejecting that which is bad, preserving and adding up all that is good; silently and insensibly working, whenever and wherever opportunity offers, at the improvement of each organic being in relation to its organic an inorganic conditions of life. (Darwin 1859:84)

Here, darkly outlined, we encounter the personification of natural selection like the Ghost of Christmas Past. "Silently and insensibly" this shadowy figure scrutinizes the natural world, "rejecting" and "preserving" without any semblance of blindness.

In fact, Darwin himself was forced to come to terms with this: "In the literal sense of the word, no doubt, natural selection is a misnomer; but whoever objected to chemists speaking of elective affinities of various elements? ...It has been said that I speak of natural selection as an active power or deity; but who objects to an author speaking of the attraction of gravity as ruling the movements of the planets?" (Peckham[Darwin] 1959:165). And in later editions of *The Origin* the initial phrase in the above passage; "It may be said...," was changed to "It may be *metaphorically* said...." (Beer 1983:69, italics added). Yet the tendency to treat Darwin's metaphor as literal truth, as we shall see, has been persistent.

THE ATTRIBUTION OF AGENCY

Essentially, Darwin's project was limited to ridding the origin of species of *metaphysical agency*. By contrast, the development of scientific explanation since the 17th century may be seen as an attempt to eliminate agency and intentionality *altogether*. Over the last 150 years, the Cartesian program in the biological sciences has tended to treat nonhuman living systems as mechanisms in which sentience is either absent or irrelevant[3] and it is perhaps no accident here that even Paley chose the analogy of the watch mechanism in his defense of divine intervention. In the context of evolutionary theory, the progressive elimination of intentionality and agency has continued for two reasons: the rejection of Lamarckism and the methodological pursuit of reductionism.

The quite reasonable repudiation of Lamarckian and other teleological explanations has led many to eschew any intentionalist account that might be held to imply the inheritance of acquired characteristics and a sense of progressive or directional evolution. In Darwin's case, this forced him into an ambivalent attitude toward intentionality, which he never resolved (Young 1985; Costall 1991).

---

[3] At times, this position has been extended to humanity, cf. Huxley (1874) and Costall's (1991) comments thereon.

However, I suggest that a concept of agency in no sense compromises a materialist account of evolution. The perceptive power of the individual, its ability to actively engage in the process of life, does not extend beyond its limited context in time and space, and hence the teleology objection dissolves. Indeed, this was a point which Darwin himself had emphasized. As Costall (1991:330–331) says, "Darwin regarded intentional activity, through its unintended effects, as the basis for the apparent purposiveness of evolution....his account was not that of blind variation at all, except in the very limited sense of blindness to very ultimate consequences."

In other words, it is central to the theory of natural selection that, by its actions, an organism *can* promote its own inherited characteristics. Moreover, it is equally possible to argue, as Baldwin (1902) did, that the actions of organisms may promote the genetic assimilation of behavioral characteristics. For example, it is widely accepted that dental reduction in late Pleistocene hominids was the result of increased reliance on technology for food processing (e.g., Brace 1979). This is a case, par excellence, of the Baldwin effect.

Meanwhile, the Cartesian method of reductionism has led to the decomposition of the organism as agent into constituent drives or traits. As Foucault (1970) observes, the traditional description of nature in terms of resemblances has progressively given way to the dissection of the world in search of the internal clockwork. Yet, it seems to me, the question of reductionism returns us directly to the analogy with artificial selection. For it is only when subjected to the power of human *agency* that an animal or plant ceases to be an agent in its own right, and only then can individual traits be identified as the objects of selection or dissection. An organism is "selected" as an integrated being because it *acts* as one. Here, then, the reductionist account of evolution compounds Darwin's mistake in his choice of the metaphor of selection.

The avenue of escape usually adopted is simply to *deny* agency in living things altogether, as Descartes had done, or to attribute a kind of Cartesian "no-nagency" to the components of organisms. This latter proposition is clearly represented in the popular writings of Richard Dawkins.

"I have emphasized that we must not think of genes as conscious, purposeful agents. Blind natural selection, however, makes them behave rather as if they were purposeful, and it has been convenient, as a shorthand, to refer to genes in the language of purpose" (Dawkins 1976:196). However, a cynic might add, in this case the language of purpose is all *too* convenient. Dawkins is in fact making a double attribution of agency; he states that "blind" natural selection "makes" genes "behave," but surely the point is that natural selection, whether blind or not, is not an agent that can compel, any more than genes are agents that can behave. Indeed, to claim that an entity can "behave...as if [it] were purposeful" without being purposeful is somewhat disingenuous.

TWO KINDS OF STRUGGLE

Returning to Darwin's own difficulties with action and intentionality, it is perhaps uncharitable to conclude, as Young (1985) does, that these were simply a weakness. As Costall (1991) suggests, it appears that Darwin was deliberately attempting to retain an intentionalist account of natural processes, and this may not be a bad idea! "Intentionality is the suppressed, if not entirely missing, link between social and biological theory, a link that may take us beyond mere analogy to a proper synthesis" (Costall 1991:331).

In my view, the root of Darwin's problem lay with the choice of the term *selection*. If, instead, one concentrates on the concept of the struggle for survival, then the problem of attribution of agency, at least in the case of natural selection, dissolves away.

As is often suggested, Darwin's theory of natural selection was strongly influenced by Malthus's writings on the struggle for existence. Yet much confusion has arisen from the fact that the concept of struggle is itself a complex construct (Bowler 1976). The key to Darwin's theory was his recognition of the role of *competition*—a struggle between organisms of the same species which is pursued through the processes of differential survival and reproduction. However, in this respect, Darwin was departing from Malthus's arguments, which largely rested on another dimension of the concept of struggle.

Malthus, although he certainly recognized the existence of competition, was essentially concerned with the relationship between limited resources and the geometric growth of human populations.[4] The Malthusian dimension of struggle, then, is the struggle for survival in relation to physical conditions of existence. As Bowler (1976) points out, this second dimension of struggle could conceivably exist in the absence of the competition:

> Even if there were a constant shortage [of resources], some other means besides competition could be employed to decide which animals and plants would live and which would die. Periodic catastrophes might simply destroy parts of the population at random, or (as the followers of Lysenko seem to have believed) the weak might simply give up to allow the strong to succeed more easily. (1976:634)

Moreover, competition is not conceived as a kind of Hobbesian "war each against all"; it is not a *direct* confrontation of "nature red in tooth and claw." Hence, the process of competition is conducted *indirectly* through the medium of the struggle for existence; organisms of the same species compete with one another through direct engagement in the physical world.

Given these distinctions, I suggest, the correct attribution of agency in the process of natural selection now becomes clear. Collectively, by their ac-

---

[4] Indeed, Malthus saw the limiting equation of resources versus population growth as the means by which the social hierarchy was maintained in a steady state.

tions, organisms are the agents of *their own* selection. In effect, the transitive verb *to select* has become intransitive at the population level. By their involvement in the *struggle for existence*, organisms are inadvertently and indirectly involved in *competition* that causes differential survival of one lineage against another. Thus, as Darwin saw, by their methodical actions, living things have an unintended effect on the process of selection.

NATURAL SELECTION: CULTURAL SELECTION

Turning to the analogy between natural and cultural processes of selection, a very real problem now comes to light. For in point of fact many models of cultural selection describe a mechanism in which human agents *select* among ideas, concepts, plans, and theories. For example: "Human choice is selective and in some ways resembles natural selection, human invention roughly resembles mutation" (Bettinger 1991:182). Or, "a naive individual uses the frequency of a variant among his models" [i.e., other people] "to evaluate the merit of a variant" (Boyd and Richerson 1985).

In a process so described, the analogy is not really with natural selection at all, but with *artificial selection*; an essentially circular argument since selective breeding is itself a form of cultural selection. By attributing the active power of decision to human individuals, this account separates the subjects of selection from the objects of selection. But in the process the fundamental force of struggle which exists in natural selection is lost. What remains is the other side of the Cartesian mind–body dualism; a metaphysically derived "free will" that directs choice. Is it possible, then, to rescue a genuinely Darwinian approach to the evolution of human culture?

Natural Selection of "Memes"

The solution adopted by Dawkins, Cloak, and others has been to attribute agency to ideas or "memes" themselves:

> As my colleague N. K. Humphrey neatly summed up an earlier draft of this chapter "…memes should be regarded as living structures, not just metaphorically but technically. When you plant a fertile meme in my mind you literally parasitize my brain, turning it into a vehicle for the meme's propagation in just the way that a virus may parasitize the genetic mechanism of a host cell." (Dawkins 1976:192)

Or:

> In a human carrier, then, a cultural instruction is more analogous to a viral or bacterial gene than to a gene of the carrier's own genome. It is like an active parasite that controls some behavior if its host. (Cloak 1975:172)

Effectively, this is not a model of cultural selection at all, but rather a natural selection of ideas themselves. Yet, in reality, both writers are simply transferring a reductionist, and wholly fallacious, biological hypothesis from genes to what Cloak (1975:168) calls the "corpuscles of culture." "Memes" can no more have volition than "genes." Indeed, it is even harder to see how "culture," which can exist in such diverse manifestations as chairs, books, buildings, or contraceptives can be said to be active (Hull 1982). The core of this fallacy is the analogy with parasites that, of course, *are* alive.[5] Its consequences, like many extreme expressions of sociobiology, can appear altogether ridiculous; "In short, 'our' cultural instructions don't work for us organisms; we work for them. At best, we are in symbiosis with them, as we are with our genes. At worst, we are their slaves" (Cloak 1975:172).

NATURAL SELECTION FOR CHOICE

Another means of explaining cultural processes is that adopted by Popper (1969); "Popper's problem of finding a force *like* natural selection to serve the function of refutation is....solved by making natural selection the moulder of the mind" (Lewontin 1982:165). This is an approach that is echoed by Alexander (1979:80):

> Cultural novelties do not replicate or spread themselves, even indirectly. They are replicated as the consequence of the behavior of the vehicles of gene replication. Only if the decision or tendencies of such vehicles of gene replication (individuals) to use or not to use a culture novelty are independent of the interests of genetic replicators can it be said that culture change is independent of the differential reproduction of genes.

This again implies that there is no cultural process of evolution at all, but simply an unfolding, predetermined by natural selection of brain architecture. This is essentially the same thing as Piaget's "Genetic Epistemology," which "cannot escape from embryological notions of developmental stages through which we must pass to adulthood, and of the unfolding of stages according to an organizational principle" (Lewontin 1982:165).

In effect, the invocation of natural selection to account for the power of decision implies that nothing is actually being transmitted culturally. The individual has a set of inherited decision rules, and the "cultural" part of the process is simply the implementation of those rules with respect to a range of choices. Thus, the Piaget/Popper model does not admit "cultural" change *per se*, and does not recognize an interaction between organism and environment

---

[5] And, of course, the controversial case of viruses, which may or may not be alive, is chosen to cloud the issue!

(Graves, 1990). For change to take place, the genetic program must, itself, be altered.

GENESIS: EPIGENESIS

Is it then the case that there is no point in talking about cultural selection as a distinct process? Is it indeed the case that the only way to develop a Darwinian approach to the past is to reduce the history of human culture to a process of natural selection? Clearly, the answer to both of these questions is no! As Bettinger points out:

> At the very best [sic],...cultural transmission sets up a complex co-evolutionary dynamic between genes and culture that cannot be reduced to the terms of a simple fitness optimizing analysis. (1991:194)

Human culture is made possible by the existence of a biological continuum. But the essential point is that our culturally based activities have a degree of independence from biological "time"; "In Darwinian theory, the sorting out of 'conjectures' is accomplished by differential birth and death rates. But, of course, children do not learn by making fatal mistakes, but by sorting hypotheses out in their heads" (Lewontin 1982:164).

The process of cultural transmission and change can operate on a shorter time scale than the cycle of birth and death. In my view, the main obstacle to explanation of this "sorting out" of "hypotheses" lies in the fact that ontogenetic process, in the period *between* birth and death, has been entirely overlooked.[6] What is needed is not a "genetic epistemology" but a form of "epigenetic naturalism" (Sinha 1984, 1988).

PROBLEMS, CHOICES, SOLUTIONS

Ontogenetically, one must admit, Piaget and Popper are at least in part correct. The solution to the problem of choice in cultural selection *begins* with the fact that, in individual development, action *precedes* choice. An infant does not choose to be born or to breathe; its initial reactions to the world are almost entirely innate. The process of becoming a human being proceeds from this basis, for, as Vygotskii (1986[1934]) argued, in order to make choices about our actions, to control and direct them, we must first be in action.

---

[6] For example, Durham (1976:94) states "Unfortunately, with our present knowledge,. it is not obvious what the various mechanisms of cultural retention are, how they function, or how they relate to one another." Thereby ignoring the past 100–150 years of psychological research!

The error of the Piagetian scheme lies in the assumption that the ability to choose is itself innate. Indeed, the implication is that the essential choices are already made and only await expression through the operations of accommodation and assimilation. But in human epigenesis, control of action is something which develops in and through a social context.

From the point of view of most cultural transmission models, the developmental process is seen as an entirely individualistic pattern of trial and error.[7] But again, this ignores the essential qualities of ontogeny. Our ability to become human beings is predicated on ontogenetic flexibility *and* its corollary of infant dependence. As social constructivists such as Vygotskii (1986[1934]) and Mead (1934) long ago realized, we *could not* develop as human beings without the support of society. Our period of maturation to adulthood is one in which we are buffered from participation in the struggle for existence; we do not die for our mistakes because our actions are not a life-and-death matter. The process of learning is thus characterized by the progressive transfer of tasks and responsibility for action *from* those around us *to us* as individuals (Rogoff and Gardner 1984; Greenfield 1984).

This process has both costs and benefits, but, contrary to the claims of those who reject "group selection" out of hand, the benefits seem to outweigh the costs.[8] As members of society, we can achieve more than we could as unaided individuals. Society presents us with what Vygotskii (1986[1934]) called a *Zone of Proximal Development*; it supports or "scaffolds" (*sensu* Wood et al. 1976) our activities; allowing us to take control of those functions that we already possess and to develop beyond our unaided capacities. As Lewontin (1982) points out, the consequences of this fact are far-reaching:

> No individual can fly by flapping her or his arms.... But people do fly as a consequence of the existence of airplanes, pilots, radios, fuel, and airports, all of which are social products.... Yet it is not "society" that flies or reads books, but individuals. In this way, social organization and individual life interpenetrate each other. Social knowledge is both the product and the producer of individual knowledge. (Lewontin 1982:168)

Human culture is something that is distributed, both in terms of the division of labor and expertise between people, and in terms of the universe of human material culture, the "airplanes, radios, fuel and airports." Our development within society guides us into participation within this network of

---

[7] Note that (contra Boyd and Richerson 1985: 81) the concept of trial and error is one which, like most choice-based models, begs the question. In order to engage in trial and error, a person or any other organism must already have a project it is attempting to accomplish—it must have criteria against which to evaluate success or failure.

[8] For more detailed discussion of the group selection argument, see Hallpike, 1986; Mayo and Gillinsky 1987; Wright 1984 and Wynne-Edwards 1986.

human and material resources, it introduces us to and equips us for the struggle for existence (Leontiev 1981). Moreover, and most importantly outside of the artificial contexts of psychology experiments and school classrooms, social scaffolding also enables us to take an active and *productive* part in everyday life. It does not present the world to us as an abstract problem to be solved, but rather guides us to participate in the completion of practical tasks whether we are individually capable of this or not (Greenfield 1984; Rogoff and Gardner 1984; Newman et al. 1984).

To give an archaeological example: In recent years, much attention has been focused on techniques of stone tool manufacture. Some writers (e.g., Davidson 1991) regard most "flint knapping" as a trial-and-error process in which useful flakes are selected from among the debris of a Piagetian-exploratory core reduction process. Yet there is growing evidence that even the earliest Acheulian and Levalloisian technologies involved complex chaînes operatoire (Boëda 1991; Perlès 1991) and if evidence from the Upper Paleolithic is to be taken as representative, such operations required an extensive period of apprenticeship (Pigeot 1990; Ploux 1991; Karlin et al. 1993). Estimates certainly seem to suggest that by the Magdalenian a period of at least 8 years study would be required to achieve mastery (Pelegrin 1991, 1993).

Quite simply, learning to make stone tools by trial and error is wasteful of time, and of potentially scarce raw material. The novice must learn to coordinate quite complex physical skills just in order to make a successful and controlled flake. More involved production techniques, such as the manufacture of blade cores or pressure flaking, require considerable time to learn. With effective guidance, the novice can participate in tasks without wasting time or raw material. This may not be all that important if the aim were simply "learning" as it is in the Piagetian paradigm. But within the practical requirements of everyday life abstract learning is a luxury.

The problem for the novice is that, to use Simon's (1969) phrase, flint knapping skills are only near-decomposable (see also Burton et al. 1984). One may identify certain subgoals within the process, but the necessary outcome, making a usable tool, is not reducible. The point, then, is that while the novice cannot produce usable tools unaided, he or she can participate in those subgoals that are within his or her grasp. For example, once the basics of knapping technique were mastered, apprentices could undertake the basic tasks of core preparation or selection of materials at quarry sites. As understanding and skill developed, so the task could be progressively transferred to the novice while the desired outcome was maintained.

More generally, then, one may see how the individual is inducted into productive activity without compromising that activity in the process. The "trial and error" model of cultural transmission implies a level of waste which is simply unacceptable in real life, most particularly where activities which involve physical risk are concerned (e.g., hunting or collecting potentially dangerous plants; see Graves 1991); while in fact the buffering role of society creates a context in which the individual may learn without risk to him- or herself, and without compromising the actions of those who are fully involved in the struggle for existence.

However, our social context also carries certain costs. To reiterate, most cultural selectionist models focus on *individual* decision-making, but in fact the individual's freedom of choice is constrained by membership of society; one is not free to do exactly as one likes. There are, of course, constraining social rules and sanctions; freedom to compete is circumscribed by interdependence in the struggle for existence, and competition, except in warfare, is never direct. But membership of society is more than a "social contract"; our very power of action is the product of social construction, we are never totally free to step outside the social constructs we are presented with—our individuality itself interpenetrates with society. Set against provision of support and scaffolding to the individual, tensions within society create obstacles to individual development and learning (Marcuse 1964; Goodnow 1989). Knowledge is deliberately withheld from us, such that the Piagetian paradigm of a world which lets the young "happily pursue their search for meaning" (Goodnow 1989:277) does not exist. In the abstract learning situation, such things as pendulums may be provided for as an aid to learning scientific principles. But in reality,

> the pendulum you are trying to understand is seldom readily available or reliable. On the contrary, it may be owned by someone else, available but rigged so that the results you get are what someone else wants you to know, or available but accompanied by the warning that experimentation might make you blind!! (Goodnow 1990:277)

What exists, then, is a continual tension within society, a dialectic driven by the centrifugal force of competition, of individual expression, against the centripetal force of social integration (Marcuse 1964; Campbell 1975; Cohen 1981). Here, social relations give rise to another form of struggle—a struggle for meaning, a struggle between social convention and individualized existence—a process which is not simply analogous to natural selection, but is effectively an expression or transformation of the basic struggle for existence. This, then, is the nexus of action for which we are seeking; the process of selection in human culture does not take place *within* us, but rather in the relations between persons—in the social dialectic.

DIALECTIC WITH THE PAST: A SHORT HISTORY OF TIME

The fundamental property which natural and cultural processes share is their historicity. In any attempt to formulate a Darwinian approach to the past, one must be aware that both the social dialectics of culture and the natural struggle for existence take place in the context of past events. Change is thus, in effect, a dialectic with the past; a renegotiation of the accumulated material of past adaptations, social practices, technologies. Anchored in the past, the dialectic is not instantaneous but extends in time, and hence the authorship of change is distributed.

All too often, this fact is forgotten; culture change is seen to be "innovative," random, springing almost from nothing. In a passage quoted earlier, Bettinger (1991:182) expresses a common assumption when he states that "human invention roughly resembles mutation." Yet, can one really say the novel practices emerge in such an unsystematic fashion? As suggested earlier, everyday experience is not a matter of abstract speculation, but of practical functionality. Whatever activity is undertaken must work in the context of engagement in the struggle for existence.

For example, apropos of the title of this paper, clocks and other devices for measuring time have a long history in human culture. Shadow clocks, water clocks, and mechanical clocks make their appearance at various times over the last 3000 years (Morgan 1980). Monuments, such as Stonehenge or Newgrange, which are aligned in terms of astronomical time, have an even longer history. In all cases, there has been a continuity of function. Each step in the process has maintained, by definition, the property of measuring time. In this sense the products of culture *are* like living organisms; unsystematic tinkering with the mechanism of a clock might easily produce a device that didn't measure time, in the same sense that a random mutation might be fatal to an organism. But, of course, the point is that people do not generally experiment at random, especially where matters of life and death are involved. The constraints of society, the scaffolding of social development, militate against this.

Again, as I have already argued for the case of living organisms, the mistake is to treat culture as a set of atomistic "ideas" or "memes," whereas, in fact, the cultural products all around us are not so constituted—they are, to repeat Simon's (1969) observation, only *near*-decomposable. So, to echo the example which Simon himself gives, the clock may be near-decomposable into subassemblies (drive mechanism, escapement and so forth), but its functioning inheres in the relationship between those sub-assemblies. In a sense, it is this relationship that gives meaning to the components of a clock—separately they are just pieces of metal, turned, milled, cast, and cut into a variety of shapes.

Consequently, the dialectics of society can only emerge from the contradictions, frictions or flaws in the integrity of cultural systems. In the case of clocks, such flaws may lie in limits of accuracy, or lack of portability, or in that, as in the case of shadow clocks and sundials, they only work during the day! But it would be wrong to assume that these contradictions exist internally to the products of culture themselves. By definition, a clock is something that measures time. The extent to which it does so satisfactorily is dictated by its function within the social matrix.

In effect, the mutationist view of cultural change is simply a variation on the old theme of technological determinism. The emergent innovation spreads through society because its superiority is transparent. Here, it is probably no coincidence that as Pfaffenberger (1988) has argued, technological determinism effectively attributes agency to artifacts themselves. For one may detect a direct parallel with the genetic/memetic determinism advocated by Cloak and Dawkins. Can an artifact, in itself, have the power to change society?

The fallacy of technological determinism, of the attribution of agency to artifacts themselves, was clearly recognized by Marx in his critique of the fetishism of commodities: "the productions of the human brain appear as independent beings endowed with life, and entering into relation both with one another and the human race" (1938:43).

The very means of production, rather than the social relations that produced them, are falsely credited with agency in social change. For, as Pfaffenberger suggests, "the Western ideology of objects renders invisible the social relations from which technology arises and in which technology is vitally embedded" (1988:242). Yet in truth cultural change is a dialectical result of the coincidence of technological possibility and social necessity.

For example, around 1580 Galileo invented the pendulum. This eventually revolutionized the accuracy of time measurement by using the force of gravity to regulate the existing verge and follet clock mechanism. Yet the first practical application of the pendulum was not developed until some 70 years later by Huygens in The Netherlands (Morgan 1980). Galileo's "experimental" approach to astronomy and physics led him to discover the regularity of the pendulum stroke. But it also brought him into direct conflict with the ideology of Christianity, in the form of the Roman Inquisition. Elsewhere, particularly in the Protestant world, an instrumental approach to science was adopted, and hence the discoveries of Copernicus, Tycho, and Galileo were adopted in spite of the fact that they challenged the geocentric/deiocentric Christian model of the universe (Burke 1978; Morgan 1980). Only where social conditions permitted could the contradictions of this model be acknowledged; the increasing desire to control time was effectively part of a shift from deiocentrism to anthropocentrism. Thus, even such an apparently simple and neutral innovation as the pendulum may be seen to depend on the coincidence of social and technical possibility. The metallurgical skills required to make clockwork mechanisms had been developed over centuries. Yet, as Kuhn (1970) suggests, purely technical "innovation" must coincide with social "revolution" for genuine and widespread change to occur—neither is alone sufficient to provoke a paradigm "shift."

CONCLUSION: PURPOSES DIVINE AND PROFANE

In truth, then, the mechanistic model of the world that we have inherited derives from analogies with artifacts and processes that, on more detailed investigation, themselves reveal the fallacy of mechanistic interpretation. The analogy of clockwork can be traced from Descartes, who equated nonhuman living things with the clockwork automata of his day, through Newton, who saw the whole universe as a deterministic clockwork set in motion by God, to Paley, who saw the clockwork complexity of life as evidence for a divine hand, a divine watchmaker:

> the watch must have had a maker.... there must have existed at some time, and at some place or other, an artificer or artificers, who formed it for the purpose which we find it actually to answer; who comprehended its construction, and designed its use. (Paley 1802)

Ironically, Paley was in part correct in his assertion; a watch must have a watchmaker; for it is not alive, is not actively involved in the struggle for existence and hence could not be the product of evolution. Yet, in truth, even the watchmaker is not totally the author of his or her creations. The purpose of the clock or watch changes with context, and in the distributed domain of human knowledge, no individual could be said to "comprehend" all aspects of its "construction." The design is not that of any individual, but the accumulated action of many watchmakers distributed through space and time.

Each age and each individual reinvents knowledge drawn from the past, much as Lock (1980) suggests that each child engages in the guided reinvention of language. This reinvention is guided and constrained by social relations, and the possibility of novelty exists only in the contradictions we each encounter through the process of reinvention. Much as Kuhn's essentially Hegelian theory suggests, the contradictions of existing theory and practice lead to new science; the contradictions of the geocentric model of the universe led Galileo to investigate gravity and to invent the pendulum. The process of cultural change, then, is one where selection emerges out of the struggle between us, the struggle for meaning and understanding within a social world.

Yet this is not to say that cultural evolution is progressive in the sense that Hegel or even Marx believed. Each new mind is engaged with a dialectic with the past, in which the future must always remain obscure. As with natural selection, the methodical activities in which humans engage always have potential for unintended consequences—the inventors of the verge and follet clock of the Middle Ages did not foresee the invention of a pendulum regulator that fitted perfectly into the escapement.

As Elias observes in the case of the "civilizing process": "obviously, individual people did not at some past time intend this change, this 'civilization,' and gradually realize it by conscious rational purposive measures. Clearly, 'civilization' is not any more than rationalization..." (1982:229). A similar point is made by Fergusson (1974:19): "Technical solutions of perceived problems have always affected a system larger than that encompassed by the planner or problem solver. Thus, in a very real sense, there is no entirely rational plan, nor can there be any assurance in the long run of a rational result." The first casualty of any endeavor is the plan! Moreover, while one may perceive a direction in the process of cultural change, this does not imply that it is in some way directed. "Directionality is in effect an artefact of evolutionary analysis. Arranging data in developmental series makes evolution look directional because the data sets have been chosen to show a full evolutionary sequence" (Cohen 1981:205).

The linear developmental series of clocks, from the sundial to the digital watch, is just one chain of nodes that can be traced through the web of human activity (cf. Borges 1970; Landau 1991). At each point one could branch off into other dialectics (e.g., the development of mechanical and electronic calculating machines) as knowledge is continually recombined in each individual's attempt to reinvent what he or she is told. In the sense that we cannot predict what use our own creations will serve for others, each of us is a blind watchmaker.

REFERENCES

Alexander, R.D., 1979, *Darwinism and Human Affairs*, University of Washington Press, Seattle.
Baldwin, J.M., 1902, *Development and Evolution*, Macmillan & Co., New York.
Beer, G., 1983, *Darwin's Plots: Evolutionary Narrative in Darwin, George Eliot and Nineteenth Century Fiction*, Ark Paperbacks, London.
Bettinger, R.L., 1991, *Hunter-Gatherers. Archaeological and Evolutionary Theory*, Plenum Press, New York.
Boëda, E., 1991, Approche de la variabilité des systèmes de production lithiquesdes industries paléolithique inférieur et moyen: Chronique d'une variabilité attendue, *Techniques et Culture* 17–18:37–79.
Borges, J.L., 1970, The Garden of Forking Paths, in: *Labyrinths*, Penguin, London, pp. 44–54.
Bowler, P.J., 1976, Malthus, Darwin and the Concept of Struggle, *Journal of the History of Ideas* 37:631–650.
Boyd, R., and Richerson, P.J., 1985, *Culture and the Evolutionary Process*, University of Chicago Press, Chicago.
Brace, C.L., 1979, Krapina "Classic Neanderthals" and the Evolution of the Human Face, *Journal of Human Evolution* 8:527–550.
Burke, J., 1978, *Connections*, Macmillan & Co., London.
Burton, R.R., Brown, J.S., and Fischer, G., 1984, Skiing as a Model of Instruction, in: *Everyday Cognition* (B. Rogoff and J. Lave, eds.), Harvard University Press, Cambridge, MA, pp. 140–150.
Campbell, D.T., 1975, On the Conflicts between Biological and Social Evolution and between Psychology and Moral Tradition, *American Psychologist* December:1103–1126.

Cloak, F.T., 1975, Is Cultural Ethology Possible?, *Human Ecology* 3:161–182.
Cohen, R., 1981, Evolutionary Epistemology and Human Values, *Current Anthropology* 22(3):201–218.
Costall, A., 1991, The "Meme" Meme, *Cultural Dynamics*, 4(3):321–335.
Darwin, C., 1859, *The Origin of Species*, Everyman, London.
Davidson, I., 1991, The Archaeology of Language Origins—A Review, *Antiquity* 65:39–48
Dawkins, R., 1976, *The Selfish Gene*, Cambridge University Press, Cambridge.
Dawkins, R., 1986, *The Blind Watchmaker*, Norton, New York.
Durham, W.H., 1976, The Adaptive Significance of Cultural Behaviour, *Human Ecology* 4:89–121.
Elias, N., 1982, *State Formation and Civilization: The Civilizing Process*, Vol. 2, Blackwell, Oxford.
Fergusson, E.S., 1974, Toward a Discipline of the History of Technology, *Technology and Culture* 15:13–30.
Foucault, M., 1970, *The Order of Things: An Archaeology of the Human Sciences*, Routledge, London.
Goodnow, J.J., 1989, The Socialization of Cognition, in: *Cultural Psychology: Essays in Comparative Human Development* (J.W. Stigler, R.A. Shweder, and S. Herdt, eds.), Cambridge University Press, Cambridge, pp. 259–286.
Graves, P.M., 1990, Sermons in Stones: An Exploration of Thomas Wynn's, "The Evolution of Spatial Competence," *Archaeological Review from Cambridge* 9(1):104–115.
Graves, P.M., 1991, The Persistence of Memory: Dynamics of Sociocultural Evolution, *Cultural Dynamics* 4(3):290–320.
Greenfield, P.M., 1984, The Theory of the Teacher in the Learning Activities of Everyday Life, in: *Everyday Cognition*, (B. Rogoff and J. Lave, eds.), Harvard University Press, Cambridge, MA, pp. 117–138.
Hallpike, C.R., 1986, *The Principles of Social Evolution*, Clarendon, Oxford.
Hull, D.L., 1982, The Naked Meme, in: *Learning, Development and Culture* (H.C. Plotkin, ed.), Wiley, New York, pp. 273–327.
Huxley, T.H., 1874, On the Hypothesis that Animals Are Automata, and Its History, *Fortnightly Review* 22:555–580.
Karlin, C., Ploux, S., Bodu, P., and Pigeot, N., 1993, Some Socio-economic Aspects of the Knapping Process Among Groups of Hunter-gatherers in the Paris Basin Area, in: *The Use of Tools by Human and Non-Human Primates* (A. Berthelet and J. Chavaillon, eds.), Clarendon, Oxford, pp. 318–340.
Kuhn, T.S., 1970, *The Structure of Scientific Revolutions*, 2nd ed., University of Chicago Press, Chicago.
Landau, M., 1991, *Narratives of Human Evolution*, Yale University Press, New York.
Leontiev, A.N., 1981, *Problems of the Development of Mind*, Progress, Moscow.
Lewontin, R.C., 1982, Organism and Environment, in: *Learning, Development and Culture* (H.C. Plotkin, ed.), Wiley, New York, pp. 151–170.
Lock, A., 1980, *The Guided Re-invention of Language*, Academic Press, New York.
Marcuse, H., 1964, *One-Dimensional Man: Studies in the Ideology of Advanced Industrial Society*, Routledge and Kegan Paul, London.
Marx, K., 1938, *Capital*, Vol. 1, Allen and Unwin, London.
Mayo, D.G., and Gilinsky, N.L., 1987, Models of Group Selection, *Philosophy of Science* 54:515–538.
Mead, G.H., 1934, *Mind, Self and Society*, University of Chicago Press, Chicago.
Morgan, C., 1980, From Sundial to Atomic Clock, in: *The Book of Time* (J. Grant, ed.), Newton-Abbot, Westbridge, pp. 84–129.
Newman, D., Griffin, P., and Cole, M., 1984, Social Constraints in Laboratory and Classroom Tasks, in: *Everyday Cognition* (B. Rogoff and J. Lave, eds.), Harvard University Press, Cambridge, MA, pp. 172–193.

Paley, W., 1802, *Natural Theology—or Evidences of the Existence and Attributes of the Deity Collected from the Appearances of Nature.*
Peckham, M. (ed.), 1959, *On the Origin of Species*, A Variorum Edition, Philadelphia.
Pelegrin, J., 1991, Les savoir-faire: une très longue histoire, *Terrain* 16:106–113.
Pelegrin, J., 1993, A Framework for Analyzing Prehistoric Stone Tool Manufacture and a Tentative Application to Some Early Stone Industries, in: *The Use of Tools by Human and Non-Human Primates* (A. Berthelet and J. Chavaillon, eds.), Clarendon, Oxford, pp. 302–317.
Perlès, C., 1991, Économie des matières premières et économie du débitage: deux conceptions opposées?, in: *25 Ans D'Études Technologiques en Préhistoire* (C. Perlès, ed.), Juan-les-Pins, Éditions APDCA, pp. 35–45.
Pfaffenberger, P., 1988, Fetishized Objects and Human Nature: Towards an Anthropology of Technology, *Man* 23:236–252.
Pigeot, N., 1990, Technical and Social Actors: Flint knapping specialists at the Magdelenian Etiolles, *Archaeological Review From Cambridge* 9(1):126–141.
Ploux, S., 1991, Technologie, Technicitétechniciens: Méthode de Détermination D'auteurs et Comportements Techniques Individuels, in: *25 Ans D'Études Technologiques en Préhistoire* (C. Perlès, ed.), Juan-les-Pins, Éditions APDCA, pp. 201–214.
Popper, K.R., 1969, *Conjectures and Refutations*, Routledge and Kegan Paul, London.
Rogoff, B., and Gardner, W., 1984, Adult Guidance of Cognitive Development, in: *Everyday Cognition* (B. Rogoff and J. Lave, eds.), Harvard University Press, Cambridge, MA, pp. 95–116.
Simon, H.A., 1969, *The Sciences of the Artificial*, MIT Press, Cambridge, MA.
Sinha, C., 1984, A Socio-naturalistic Approach to Human Development, in: *Beyond Neo-Darwinism* (M.W. Ho and P.J. Saunders, ed.), Academic Press, New York, pp. 331–362.
Sinha, C., 1988, *Language and Representation*, Harvester Wheatsheaf, Hemel Hempstead.
Vygotskii, L.S., 1986 [1934], *Thought and Language*, MIT Press, Cambridge, MA.
Wood, D.J., Bruner, J.S., and Ross, G., 1976, The Role of Tutoring in Problem Solving, *Journal of Child Psychology and Psychiatry* 17:89–101.
Wright, S., 1984, Theories of Group Selection, in: *Genes, Organisms, Populations. Controversies over the Units of Selection* (R.N. Brandon and R.M. Burian, eds.), MIT Press, Boston, pp. 40–41.
Wynne-Edwards, V.C., 1986, *Evolution by Group Selection*, Blackwell, Oxford.
Young, R.M., 1985, *Darwin's Metaphor: Nature's Place in Victorian Culture*, Cambridge University Press, Cambridge.

PART IV

COGNITION AND THE EVOLUTION OF MENTAL ADAPTATIONS

The view that the human mind is a tabula rasa whose content is determined by group-level processes has played a central role in anthropological theory, but has come to be increasingly discredited by new advances in evolutionary biology and cognitive science. The decline of this view, and its replacement by more evolutionarily and psychologically realistic alternatives requires restructuring certain widely held views in archaeology. New views of our evolved species-typical psychological architecture indicate that it is not well characterized as a passive and content-free "capacity for culture," but instead has a complex organization that imposes its own order on the social world and on subsistence practices. This architecture consists of a collection of information-processing adaptations that evolved during the Pleistocene, and which are structured to solve specific families of adaptive problems, such as foraging, effort allocation, mate choice, mate defense, social exchange, inbreeding avoidance, resource defense, aggressive threat, coalition formation, warfare, status and power competition, and so on. These developments make the study of evolutionary psychology and archaeology relevant to each other. The known behavior of each human group provides data that can be used to evaluate the adequacy of the nature of our evolved species-typical psychology. Reciprocally, an increasingly detailed model of the human psychological architecture can serve as a powerful new tool for archaeologists. Complex psychological and physiological adaptations appear to be species-typical, and so by virtue of the fact that archaeologists are studying humans, they know that whatever recent humans they are studying manifested this array of complex behavior-regulatory adaptations, regardless of the existence of cultural variation. Moreover, knowledge of these psychological adaptations fills in missing links in hypotheses about how ecological actors regulate behavior. Finally, archaeologists will need to abandon the notion that groups are themselves adaptive systems in the traditional sense. Instead, group-level phenomena are generated by the dynamics that emerge from individual decision-makers interacting in populations. These individuals are guided by psychological programs that evolved to bring about outcomes that in the Pleistocene would have promoted the inclusive fitness of the decision-maker.

Chapter 10

Weak Modularity and the Evolution of Human Social Behavior

JAMES STEELE

INTRODUCTION

Cognition is one of the meeting points of genes and culture, and understanding human cognitive evolution is therefore central to our understanding of the evolution of human cultural capacities. Currently this is hampered by the polarization of debate on "universals" and cognitive modularity. The straw man options have been Model 1 ("general intelligence")—human cognition entails content-independent, domain-general generative and analytical skills, generalized across domains such as language, object-manipulation, and social interaction—and Model 2 ("strong modularity")—human cognition entails genetically determined domain-specific abilities which are not transferable across domains. Model 1 is generally wedded to encephalization models of human brain evolution, and tends to be associated with a long time frame, gradualism, and an emphasis on similarities between human and nonhuman cognition. The paradigm is most strongly associated with Piaget and his school in cognitive-developmental psychology. Model 2 is generally wedded to reorganization models of human brain evolution, and tends to be associated with a telescoping of the time frame to emphasize a late, differentiating "last spurt" in human cog-

JAMES STEELE • Department of Archaeology, University of Southampton, Highfield, Southampton SO17 1BJ, United Kingdom.

nitive evolution in adaptation to Pleistocene environments. It is most closely associated with Chomsky's position in linguistics, and with Fodor's work in cognitive science.

The weaknesses of these two positions are well known. Model 1 fits well with the pattern of human brain evolution (expansion of the neocortex), but not with the developmental canalization of language skills. Model 2 fits well with the observed domain bias in language acquisition, but not with the pattern of human brain evolution (in which a strong language module is not the salient evolutionary novelty).

In this paper I argue for a third option. Model 3 ("weak modularity") recognizes the existence of dedicated circuitry in the human brain which predisposes to the development of certain skills and abilities. It also recognizes the necessity of a "normal social environment" for their full development in any individual. The concept of a "normal social environment" in turn implies that there are aspects of human social systems which are highly canalized, adaptive, and constant across cultures. In the next section I show how language and social exchange develop as "weak modules," and in the third section I develop some suggestions concerning the evolution of the "normal social environment."

LANGUAGE AND COOPERATION AS WEAKLY MODULARIZED ABILITIES

Dedicated circuitry exists in the human neocortex for language and for social cognition. However, the basic dedicated circuits are evidently conserved and shared with nonhuman primates. Two aspects are novel: the coupling of these circuits with greatly expanded memory capacity, both for long-term memory and for executive or "working memory"; and the coupling of these brain systems to a modified peripheral skeleto-muscular apparatus (the vocal tract in the case of language).

The existence of conserved, lateralized circuitry linking the Broca and Wernicke area homologues in macaques has been demonstrated by Deacon (1988a), and is also implied by work on the comprehension disorders associated with localized left temporal lobe lesions by Heffner and Heffner (1989). It is often argued that since the Broca's area homologue has no apparent involvement in monkey vocalizations, its role in human speech is exaptive. However, the dichotomy between the species is not complete: Sutton and Jurgens (1988) discuss findings of vocal chord movement and vocalization resulting from electric stimulation of this area in chimpanzees and macaques (Sutton and Jurgens 1988:637), and note that future work with nonhuman primates may show involvement of the lateral neocortex in complex vocal output where simple signals are "chained" into strings, as in gibbon "song." Thus, the dedicated circuitry in human speech processing, which makes speech "modular" in the strong sense, is probably a conserved feature.

The same applies to social cognitive circuitry. Brothers and Ring (1992) note that social cognition appears to be anatomically modular, in monkeys and humans alike, and that such a high-level module may be based on a number of different classes of input, including faces, voices, and visual and auditory affect. They suggest that "evolutionarily speaking, there should be a premium on the accurate assignment of qualities such as 'helpful', 'generous', 'selfish', and 'untrustworthy' to other individuals" (Brothers and Ring 1992:115). Monkey face processing and facial emotional expressivity is lateralized to the right hemisphere, as in humans: in both, there is a face-specific area in the temporal cortex, and in both, there is a nerve fiber bundle radiating from the occipital lobe into the temporal lobe, damage to which can present in humans as a deficit in recognition of familiar faces (prosopagnosia) (Tusa and Ungerleider 1985, DeSimone 1991). Hauser (1993) has found that macaques also share the human pattern of right hemisphere dominance for facial emotional expression, just as they share the human pattern of left hemisphere dominance for perception of same-species vocal signals (Petersen et al. 1978). Thus, major components of human social cognition are also conserved and shared with other anthropoid primates. Povinelli (1993) suggests that awareness of the existence of mental states, in oneself and in others, is probably observable in great apes, but not in monkeys. Such "metacognition"—the attribution of causal force to mental states—is essential to human social agency. Povinelli suggests that evolved differences in ontogeny underlie this discontinuity. This complements work by Parker and Gibson (e.g., 1979) on the relationship of brain size to cognitive development in primates: but the underlying modular architecture is likely based on the dedicated circuitry which can also be found in rhesus monkeys.

What differentiates human brains from those of other primates is the size of the neocortex, and the proportion of it which is given over to the working memory functions which underlie complex behavior. It is disputed whether this reflects an effect of having a brain which is larger in absolute terms (Passingham 1982), or whether this reflects some gross pattern of cerebral reorganization (Deacon 1988b). What is agreed, however, is that this quantitative difference has a qualitative effect on the processing of specific, "wired up" complex behavior patterns such as speech production and comprehension, and cooperative reasoning.

In summary, human language and social cognitive abilities have some of the characteristics of "strong modules." However, the dedicated circuitry is shared with other primates. Furthermore, these dedicated circuits are not sufficient in themselves to cause the appearance of functional language and social behavior in any individual. In cases such as that of Genie, an adolescent found reared in extreme social isolation who subsequently acquired good lexical and propositional semantic skills, but with persistent grammatical and pragmatic impairment (Curtiss 1988), the pragmatic deficit must relate to the lack of a "normal social environment" for early language development. In fact, Locke (1993) has argued that social cognitive skills precede the appearance of a Grammar Acquisition Module both phylogenetically and ontogenetically. He argues that functional linguistic competence presupposes sensitivity to social contexts of use, and that these aspects of communication are inferred by the infant from paralinguistic aspects of adult speech addressed to him or her before the development of grammatical ability (using a specialized capacity for social cognition lateralized to the right hemisphere).

A similar case can be made for the development of cooperative reasoning. Cosmides (1989) has shown how human conditional reasoning skills vary with content—refuting the Model 1 assumption that we reason with a content-independent propositional reasoning strategy. She found that where the content related to social contracts, conditional reasoning was most facilitated. The inference made was that humans have a strong (Model 2) bias to reason effectively about social exchange rules, and that this reflects the importance of social cooperation in the social environment of Pleistocene human foragers.

I recently replicated this finding with a small sample of undergraduates (n=33, 18 male, 15 female, all anonymous volunteers) (Steele 1993). Where the conditional reasoning problem was phrased as a social contract, the overall success rate in solving it was 61%. Where the problem was phrased in descriptive (nonsocial contract) terms, the solution rate ran at 11%. What was interesting, however, was that not everyone fared equally well with the social contract versions of the problem. The students had also filled out a personality questionnaire (Christie and Geis 1970) and completed a simple "IQ-type" test (Shipley 1940). The personality questionnaire, which related to "Machiavellianism" in personal ethics, was of 20 years' vintage, and somewhat gender-biased in its phrasing. Intriguingly, the only four male subjects to get all three social contract versions of the problem wrong were the four who were most "Machiavellian" (Figure 1). All four were well above the norm for this trait, and thus can be described as abnormally manipulative, hostile, and alienated in their outlook on human motivations and personal relationships. Incidentally, these four were completely normal in their scores on abstract and verbal reasoning as gauged by the brief "IQ-type" test.

My tentative interpretation of this result (from a very small pilot study) is that the facilitation which Cosmides observed across her Harvard undergraduate sample when reasoning about social rules does not reflect a strong (Model 2) cognitive module. Rather, we are looking at a weak (Model 3) module for social exchange reasoning that develops with the experience of a "normal social environment." The four male undergraduates in my sample who were hyper-

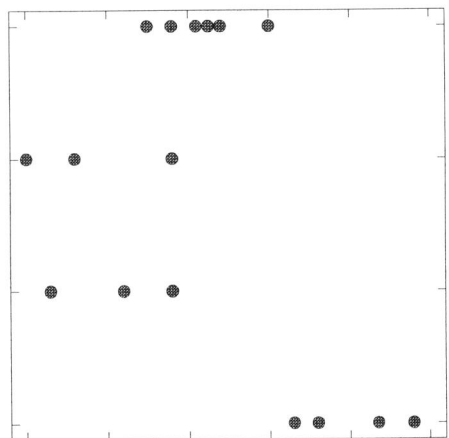

Figure 1. Effect of Machiavellianism on reasoning about social exchange. Y-variable is number of correct solutions to social contract versions of the Wason task, X-variable is Mach V score (average U.K. student score was previously found to be 99, Rogers and Semin 1973). $n = 18$ male undergraduates. Data from Steele 1993. The horizontal bar gives the mean and one standard deviation for scores on the Mach V test in a much larger British college-age sample (99 ± 12), from Rogers and Semin 1973.

Machiavellian, and who showed no facilitation in reasoning when the problem concerned social rules, had—I guess—been in some way "held back" in their cognitive-emotional development in this regard. Either that, or they represent an alternative, frequency-dependent social strategy similar to the "always defect" strategy in game theory (Wilson *et al.* 1996).

The implication of such work is that language development and the development of cooperative reasoning are not so strongly buffered as to be impervious to abnormalities in the developing individual's social environment. Rather, as extreme cases such as Genie, or as the less extreme variation of the hyper-Machiavellian males in my classroom sample show, dedicated circuitry requires the buffering of a "normal social environment" in which to develop. This is the basis of Model 3—"weak modularity."

HUMAN COGNITIVE EVOLUTION AND THE "NORMAL SOCIAL ENVIRONMENT"

Proponents of Model 2 modularity such as Cosmides (1989) tend to associate human cognitive evolution with inferred social adaptations of the Pleistocene. Such associations are rarely developed into archaeologically testable quantitative hypotheses.

For the past 15 years the principal archaeological model of the evolution of the human "normal social environment" has been Isaac's "central place foraging" model, based on the sexual division of provisioning labor in the nuclear family group. Isaac saw this system as emerging with early *Homo*, with the basic human foraging and settlement pattern continuing through the period of evolution of our genus (e.g., Isaac 1978; Kroll and Isaac 1984).

More recently, however, an alternative model of the evolving human "normal social environment" has emerged based on the assumption that group (band) size is more important than the nuclear unit for understanding hominid social systems, and that mean group size increased during the evolution of our genus. A Darwinian inference is that with a socioecological niche promoting larger groups with a lower mean coefficient of relatedness between members, there were increased fitness payoffs to parents who invested more time in socializing their offspring to be effective social agents. The human "normal social environment" is then one of intensified parental investment in their offspring's social learning, with content primarily oriented to pragmatic social reasoning and communication skills.

Examples of work in this vein include Dunbar's (1993) arguments from primate socioecology for a link between hominid encephalization (genus *Homo*), language, and increased group size, and Mithen's (1995) arguments that the Acheulian biface tradition appears in the Lower Paleolithic record only during climatic phases when natural resource structure would have increased predation risk, and thus boosted group size and persistence above the threshold for the establishment of a "tradition." Within the later hominid line, arguments for an evolution of social systems from *Homo erectus* to *Homo sapiens sapiens* have been based on the increased incidence of symbolic and artistic behaviors in the Upper Paleolithic record of Europe, although these have not generally been linked to quantitative parameters like group size. But as Binford (1989:37) puts it, "as archaeologists we will not grasp the transition unless we begin to worry about how to measure variables such as planning depth, mobility, group size, and compositional variability and then proceed to see how these properties vary with environmental conditions as a clue to the ecology of ancient populations."

How then can we make archaeologically falsifiable predictions about the nature of hominid "normal social environments"? We need to avoid concentrating on variation and neglecting the information inherent in a central tendency. Primatologists are used to comparing species by their mean scores for behavioral and anatomical traits (such as group size and brain size). This does not mean that they ignore the variation: primate group size variability is an important topic in its own right (Melnick and Pearl 1987; Dunbar 1988; Beauchamp and Cabana 1989). However, quantitative comparison of primate taxa using mean values for such traits has provided many insights into primate physiological ecology, and the relationship between anatomy and social systems. This work provides a quantitative basis for predicting variation in primate—including human—social systems along common dimensions (in contrast to Isaac's "phylogenetic" technique of tabulating derived human behavioral traits and looking for archaeological evidence of the branching points at which they first appeared).

Recently, two anthropologists with backgrounds in primate studies have argued that increases in group size were crucial to human evolution, and that the group sizes of extinct hominids can be predicted from the sizes of their brains (Aiello and Dunbar 1993). Aiello and Dunbar think that group size can be predicted from the amount of neocortex a primate has in its brain, relative to other brain structures. The argument is that this reflects the computational load of social relationships in complex networks (where the number of potential relationships increases as a parabolic function of the number of individuals in the network). Thus, we are presented with a set of predicted group sizes for various fossil hominids with brains of known endocranial volume, based on a best-fit line previously calculated for the nonhuman primates (in Dunbar 1992).

This approach needs to be further developed. The same authors (Aiello and Dunbar 1993:189) predict from brain size data that orangutans should live in groups of 60 individuals, and gorillas in groups of 67—yet the mean group size for orangs is two individuals, one female and usually only one dependent offspring (Mackinnon 1974), while for gorillas the median group size is seven individuals (Harcourt et al. 1981). The proportion of neocortex increases with absolute brain size across anthropoid genera: bigger brains have a higher neocortex ratio (Figure 2). Comparing primate genera, we find that big bodies tend to go with bigger brains, other things being equal. Thus, orangs and gorillas have high neocortex ratios by virtue of being big animals, not by virtue of having to cope with large group sizes (see also Steele 1996).

We should not, however, throw out the baby with this bath water. If we look at residual variation in neocortex ratio across anthropoid primate genera, after controlling for body size, we see that there is a residual correlation with

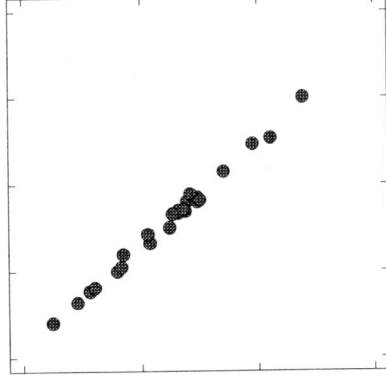

Figure 2. Neocortex plotted against volume of the rest of brain (both mm^3) for selected anthropoid species ($n=27$). Data from Stephan et al. (1981). $r^2 = 99.3$, $p < 0.0001$. Least-squares regression model: log neocortex volume (Y) = 1.19 log rest of brain volume (X)—1.11. The slope indicates that a ratio of neocortex to rest of brain volume will increase with total brain size.

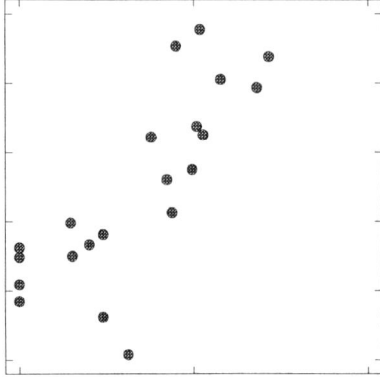

Figure 3. Residual correlation between neocortex ratio (controlled for body weight) and group size (mean number of adult females), the "X" variable, for selected anthropoid genera (n=21). Ratio is log neocortex/log rest of brain volume (data from Stephan et al. 1981). Residuals are derived from a least-squares regression of this ratio against log body weight. Generic mean group size data (no. of adult females) are from Smuts et al. (1987), and apply only to species represented in the brain data tables. Data compilation in Steele (1993). $r^2 = 62.1$, $p < 0.001$.

the mean number of adult females in a group (Figure 3). In other words, while animals with big bodies and correspondingly big brains will also have high neocortex ratios, there is an additional tendency for residual bias toward the neocortex to associate with living in larger social groups. The assumption that this is a functional and adaptive relationship, reflecting computational constraints on social complexity, remains as yet unchallenged.

All this is relevant to the evolution of the human "normal social environment." McHenry (1988) has estimated that total brain size, controlled for body size, did not increase from *Homo habilis* to *Homo erectus*. If African *Homo erectus* was smarter, this was simply because a bigger-bodied species took a correlated hike in absolute brain size, with entailments for the neocortex ratio. Conventional archaeological thinking is that before the Late Pleistocene, *Homo* populations were of low density, widely dispersed, and with little organization above the band level (and see Steele 1994). By contrast, *Homo sapiens sapiens* has not just a bigger brain in absolute terms, but also an increased relative brain size, the pattern we expect as a correlate of living in larger groups. Modern humans also show evidence of anatomical adaptations for speech (vocal tract remodeling) (Lieberman 1984), which Dunbar's (1993) "vocal grooming" model of language origins would link to increased group sizes.

These observations and expectations, derived from comparing living primates and from analyzing the fossil and cultural evidence of hominid evolution, converge on a representation of the evolved "normal social environment" of *Homo sapiens sapiens* in which social interactions are relatively complex, and the number of relationships relatively large compared to the pattern of the Lower Paleolithic. The pattern of weakly modular cognitive evolution seen in humans implies the coevolution of a linguistically dense social environment sufficiently complex to sustain a "distributed culture" based on cooperative relations. The fossil and archaeological evidence suggests that this social environment appeared quite late in the course of human evolution. However, we need now to develop ways of quantifying variation in the archaeological record relevant to quantitative parameters such as mean group size and home range area, so that work on the Paleolithic record can generate input into broad comparative models of the kind being developed by Aiello and Dunbar.

CONCLUSIONS

One of the appeals of a Darwinian approach to archaeological interpretation is that it might foster a new family of simple, deterministic models of human behavior. This is in the context of an ascendant intellectual movement—postmodernism—which is built on the concepts of indeterminacy and of context-dependent meaning.

We might seek to increase the constraints on interpretation by increasing the number and power of the predispositions or universals which are held to bias human behavior in predictable directions. Human sociobiology has been a fertile source of conjecture in this area. However, the recent rise of "evolutionary psychology," which attempts to fit predispositions and biases into their adaptive contexts in Paleolithic societies, may take this trend too far. As I have shown in this paper, behavior which may appear as a strong derived module with no homologues in other species can, when examined closely as to the nature of the mechanism, prove both to be in essence conserved and to be only weakly modularized.

Darwinism in archaeology means many things, but not least among them is the drive to describe and explain central tendencies and ranges of variation for human behavioral traits across regions, periods, and economic systems. In this paper, I have argued for the need to find archaeological data on mean and variation in hominid group sizes. This is necessary to test any explanation of human cognitive evolution which (like mine) posits "weak modules," and invokes a "normal social environment" characterized by large social groups. The current contrast between two representations of the human "normal social environment"—Isaac's model of central place foraging, and the group size models proposed by Aiello and Dunbar, and others—can be resolved into a difference over the scale of hominid social systems, and the extent to which that scale was constant over the course of evolution of *Homo*. As Darwinian archaeologists, we should be well placed to resolve such disputes.

ACKNOWLEDGMENTS

I am very grateful to Herb Maschner and Steven Mithen for inviting my participation in this volume.

REFERENCES

Aiello, L.C., and Dunbar, R.I.M., 1993, Neocortex Size, Group Size, and the Evolution of Language, *Current Anthropology* 34:184–193.

Beauchamp, G., and Cabana, G., 1989, Group Size Variability in Primates, *Primates* 31:171–182.

Binford, L.R., 1989, Isolating the Transition to Cultural Adaptations: An Organizational Approach, in: *The Emergence of Modern Humans* (E. Trinkaus, ed.), Cambridge University Press, Cambridge, pp. 18–41.

Brothers, L., and Ring, B., 1992, A Neuroethological Framework for the Representation of Minds, *Journal of Cognitive Neuroscience* 4:107–118.

Christie, R., and Geis, F.L. (eds.), 1970, *Studies in Machiavellianism*, Academic Press, New York.

Cosmides, L., 1989, The Logic of Social Exchange: Has Natural Selection Shaped How Humans Reason? Studies with the Wason Selection Task, *Cognition* 31:187–276.

Curtiss, S., 1988, Abnormal Language Acquisition and Grammar: Evidence for the Modularity of Language, in: *Language, Speech and Mind* (L.M. Hyman and C.N. Li, eds.), Routledge, London, pp. 81–102.

Deacon, T.W., 1988a, Human Brain Evolution: I. Evolution of Language Circuits, in: *Intelligence and Evolutionary Biology* (H. Jerison and I. Jerison, eds.), Springer, Berlin, pp. 363–382.

Deacon, T.W., 1988b, Human Brain Evolution: II. Embryology and Brain Allometry, in: *Intelligence and Evolutionary Biology* (H. Jerison and I. Jerison, eds.), Springer, Berlin, pp. 383–416.

DeSimone, R., 1991, Face-selective Cells in the Temporal Cortex of Monkeys, *Journal of Cognitive Neuroscience* 3:1–8.

Dunbar, R.I.M., 1988, *Primate Social Systems*, Croom Helm, London.

Dunbar, R.I.M., 1992, Neocortex Size as a Constraint on Group Size in Primates, *Journal of Human Evolution* 20:469–493.

Dunbar, R.I.M., 1993, Coevolution of Neocortex Size, Group Size and Language in Humans, *Behavioural and Brain Sciences* 16:681–735.

Harcourt, A.H., Fossey, D., and Sabater-Pi, J., 1981, Demography of *Gorilla gorilla*, *Journal of Zoology (London)* 195:215–233.

Hauser, M.D., 1993, Right Hemisphere Dominance for the Production of Facial Expressions in Monkeys, *Science* 261:475–477.

Heffner, H.E., and Heffner, R.S., 1989, Cortical Deafness Cannot Account for the Inability of Japanese Macaques to Discriminate Species-specific Vocalizations, *Brain and Language* 36:275–285.

Isaac, G.L., 1978, The Food-sharing Behavior of Protohuman Hominids, *Scientific American* 238:90–108.

Kroll, E.M., and Isaac, G.L., 1984, Configurations of Artifacts and Bones at Early Pleistocene Sites in East Africa, in: *Intrasite Spatial Analysis in Archaeology* (H. Hietala, ed.), Cambridge University Press, Cambridge, pp. 4–31.

Lieberman, P., 1984, *The Biology and Evolution of Language*, Harvard University Press, Cambridge, MA.

Locke, J.L., 1993, *The Child's Path to Spoken Language*, Harvard University Press, Cambridge, MA.

McHenry, H.M., 1988, New Estimates of Body Weight in Early Hominids and Their Significance to Encephalization and Megadontia in 'Robust' Australopithecines, in: *Evolutionary History of the 'Robust' Australopithecines* (F.E. Grine, ed.), Aldine de Gruyter, New York, pp. 133–148.

Mackinnon, J., 1974, The Behavior and Ecology of Wild Orang-utans (*Pongo pygmaeus*), *Animal Behavior* 22:3–74.

Melnick, D.J., and Pearl, M.C., 1987, Cercopithecines in Multimale Groups: Genetic Diversity and Population Structure, in: *Primate Societies* (B.B. Smuts, D.L. Cheney, R.M. Seyfarth, R.W. Wrangham, and T.T. Struhsaker, eds.), University of Chicago Press, Chicago, pp. 121–134.

Mithen, S.J., 1996, Social Learning and Cultural Tradition: Interpreting Early Palaeolithic Technology, in: *The Archaeology of Human Ancestry* (J. Steele and S. Shennan, eds.), Routledge, London, pp. 207–229.

Parker, S.T., and Gibson, K.R., 1979, A Developmental Model for the Evolution of Language and Intelligence in Early Hominids, *Behavioral and Brain Sciences* 2:367–408.

Passingham, R.E., 1982, *The Human Primate*, Freeman, San Francisco.

Petersen, M.R., Beecher, M.D., Zoloth, S.R., Moody, D.B., and Stebbins, W.C., 1978, Neural Lateralization of Species-Specific Vocalizations by Japanese Macaques (*Macaca fuscata*), *Science* 202:324–327.

Povinelli, D.J., 1993, Reconstructing the Evolution of Mind, *American Psychologist* 48:493–509.

Rogers, R.S., and Semin, G.R., 1973, Mach V: An Improved Scoring System Based on a Triadic Choice Model, *Journal of Personality and Social Psychology* 27:34–40.

Shipley, W.C., 1940, A Self-administering Scale for Measuring Intellectual Impairment and Deterioration, *Journal of Psychology* 9:371.

Smuts, B.B., Cheney, D.L., Seyfarth, R.M., Wrangham, R.W., and Struhsaker, T.T. (eds.), 1987, *Primate Societies*, University of Chicago Press, Chicago.

Steele, J., 1993, *Primate Perspectives on Human Evolution*, Unpublished Ph.D. thesis, University of Southampton, England.

Steele, J., 1994, Communication Networks and Dispersal Patterns in Human Evolution: A Simple Simulation Model, *World Archaeology* 26:126–143.

Steele, J., 1996, On Predicting Hominid Group Sizes, PHGS, in: *The Archaeology of Human Ancestry* (J. Steele and S. Shennan, eds.), London: Routledge, pp. 230–252.

Stephan, H., Frahm, H., and Baron, G., 1981, New and Revised Data on Volumes of Brain Structures in Insectivores and Primates, *Folia Primatologia* 35:1–29.

Sutton, D., and Jurgens, U., 1988, Neural Control of Vocalization, in: *Comparative Primate Biology*, Vol. 4 (H.D. Steklis and J. Erwin, eds.), Liss, New York, pp. 625–648.

Tusa, R.J., and Ungerleider, L.G., 1985, The Inferior Longitudinal Fasciculus: A Reexamination in Humans and Monkeys, *Annals of Neurology* 18:583–591.

Wilson, D.S., Near, D., and Miller, R.R., 1996, Machiavellianism: A Synthesis of the Evolutionary and Psychological Literatures, *Psychological Bulletin* 119:285–299.

Chapter 11

The Origin of Art
Natural Signs, Mental Modularity, and Visual Symbolism

STEVEN MITHEN

INTRODUCTION

Recent attempts to explain the origin of visual symbolism have focused on image making (Davis 1986), and the relationships between perception, depiction, and language (Davidson and Noble 1989). I build on this work by developing a two-stage model for the evolution of visual symbolism. The first stage concerns the evolution of the ability to attribute meaning to visual images. For this I consider the role of 'natural signs'—epitomized by the tracks and trails left unintentionally by animals. The significance of these has been prematurely dismissed by Davidson and Noble (1989). The second stage concerns the integration of this ability, with those for making marks, communicating intentionally, and classifying signs. Taken together, these four physical/cognitive processes constitute the "capacity for visual symbolism." I discuss the process of their integration by drawing on recent work concerning mental modularity, accessibility, and hierarchization in cognitive evolution. In essence I concur with Mellars (1991) that the Middle/Upper Paleolithic transition, of which the appearance of visual symbolism is a fundamental feature, marks a critical threshold in cognitive evolution. But, rather than invoking language, which is more likely to have arisen much earlier in evolution, the major event is likely

---

STEVEN MITHEN • Department of Archaeology, University of Reading, Whiteknights, P.O. Box 218, Reading RG6 2AA, United Kingdom.

to have been increased accessibility between mental modules allowing the development of high-level cognitive processes.

THE FIRST ART

The earliest representational art appears 32,000 years ago in the form of carved animals and human figures from Vogelherd, Geissenklösterle, and Hohlenstein-Stadel in southern Germany (Hahn 1972, 1984). At a similar date in western France, ivory beads were carved to represent seashells, as found at La Souquette (White 1989). The representational nature of these images suggests that they are likely to have been symbolic (Davis 1986). In both of these regions one finds images repeatedly engraved on stone surfaces which are likely to have been noniconic symbols. Most notable are the "vulva" signs from La Ferrasie and nearby sites (Delluc and Delluc 1978) in southwest France. Aurignacian contexts have also provided a diverse range of artifacts for personal decoration such as beads, pendants and perforated animal teeth, which may also have carried symbolic information (White 1989).

Prior to 32,000 B.P. numerous artifacts have been recovered with marks and ochre stains. These have been reviewed by Marshack (1990) and Bednarik (1992), both of whom believe this indicates that Lower and Middle Paleolithic hominids had the capacity to create visual symbols. However, many, if not all, of these marks may have been created unintentionally in the course of other activities, such as cutting plant material on a bone support (Chase and Dibble 1987, 1992; Davidson 1990, 1992). Even if they were intentional markings, they may have been "self-sufficient," with no semantic content (Davis 1986). The most convincing evidence for some intentional marking prior to 32,000 B.P. is the presence of manganese oxide and red ochre in Middle and Lower Paleolithic contexts (Wreschner 1980; Marshack 1981). The character of these suggests that color was being applied to something, but again the symbolic character of the patches of color created is unknown (Mellars 1989).

The principal argument against visual symbolism prior to the Upper Paleolithic is the uniqueness of each individual mark—there is no evidence for the repeated use of similar images. In the absence of such patterning, it is unlikely that these marks carried symbolic meanings (Chase 1991). Duff *et al.* (1992) question this argument, pointing to the extremely poor preservation of the early Paleolithic archaeological record. In addition, they stress the contrast between the possession and the expression of a behavioral capacity arguing that the ability to produce visual symbols was present in hominids from the early Pleistocene, but found little expression until the later Upper Paleolithic. This is not a very convincing argument. As Wobst (1977) suggested, it is likely that once something symbolic had been created, everything had the potential to be symbols, and hence this capacity could not be 'switched on and off in the manner Duff et al. suggest. My own position follows that of Chase and Dibble (1987) and Davidson (1990) in failing to be convinced by the evidence that visual symbols were created by Middle and Lower Paleolithic populations.

SYMBOLS AND THE CAPACITY FOR VISUAL SYMBOLISM

The images created in certain areas after 32,000 B.P. are likely to have shared the characteristics of visual symbols created by ethnographically documented hunter-gatherers. While the definition of a "symbol" is notoriously difficult, at least five properties are critical:

1. The form of a symbol may be arbitrary to its referent. Also, when a symbol is representational, that which is depicted may not necessarily be its referent, e.g., a painting of a bison may mean a horse.
2. There may be considerable space/time displacement between the symbol and the object or event to which it refers.
3. The meaning of a symbol is shared between at least two individuals.
4. While general meanings are shared, they may also vary according to the knowledge of the observer;
5. The form of a symbol may be unique, but nevertheless correctly interpreted by an observer without having had prior experience of the specific symbol. This is most likely to arise when symbols are unique compositions of separate elements, each of which may have a distinct referent. Also, the same symbol may tolerate some degree of variability, either imposed consciously by its creator (e.g., a particular style) or arising from technical constraints on making exact copies.

These properties of symbols are particularly apparent when ethnographic data are available to interpret the rock art of hunter-gatherers. For instance, Bahn (1986) has described how similar images in aboriginal rock art may have very different types of referents, some of which bear a visual resemblance to the symbol. Many aboriginal paintings have also been shown to carry multiple meanings ranging from the "outside," most literal interpretations to the "inside" meanings which refer to the actions of Ancestral beings (Morphy 1989; Taylor 1989). Knowledge of the inside meanings may only be revealed following initiation ceremonies. Moreover, both the inside and outside meanings may refer to events displaced in space and time from the rock art site, such as a distant water hole. Many of the rock paintings consist of unique compositions of elements which are created and read as a whole although that particular combination may not have been encountered. Studies of individual artists have shown how some traditions, e.g., X-ray style, can tolerate considerable individual variability and creativity (e.g., Haskovec and Sullivan 1989). It is likely that these properties of symbols are universal and apply to those created during the Paleolithic.

The ability to create and 'read' such visual symbols requires at least four physical and cognitive processes:

1. The making of visual images.
2. The identification that a particular image belongs to a certain class of symbol, or constitutes the first member of a new class.
3. Intentional communication with reference to some displaced event or object.
4. The attribution of meaning to a visual image not associated with its referent.

To account for the evolution of the "capacity for visual symbolism" we will need to account for the evolution of each of these abilities as well as the manner in which they become integrated.

THE EVOLUTION OF VISUAL SYMBOLISM

Davis (1986) discussed the origin of image making, claiming that "image making originated in the discovery of the representational capacity of lines, marks and blots of color which need not, and often do not, have a representational status" (1986:194). This is a truism—how could it have been otherwise? As Layton (1986) remarked, we need to develop hypotheses for what natural selection pressure or social function would favor the development of the necessary cognitive and manual skills: "why should a blot be taken for a hole or a mark for a thing" (Layton 1986:205). Similarly, Duff et al. (1992) have argued that the important question is to identify the conditions that would have selected for an increasing symbolic component to human behavior at particular times and places. I concur with Duff et al. that symbolism probably developed in contexts distinct from those in which it is primarily manifest today. Specifically, I will argue that the critical component of symbolic behavior—the attribution of meaning— arose in relation to the physical environment, whereas, once evolved, visual symbolism principally operates in the social sphere.

According to Davidson and Noble (1989), the attribution of meaning to marks could only have taken place after the evolution of language since they adhere to the notion that human perception is mediated by language. In a somewhat circular, if not contradictory, argument, they suggest that language could only have arisen from depiction—the discovery that signs could be used to convey meaning. This, they argued, may have arisen from the accidental "freezing" of a gesture as a visual image (a line in sand or mud) which would then have been seen to be associated with the gesture and perceived to convey the same meaning. This scenario appears to require that at least two individuals have this realization simultaneously—for it is the essence of visual symbols that meaning is shared.

This raises a fundamental problem with explaining the origin of visual symbols. It is part of the definition of a symbol that its meaning is shared by at least two individuals. Even if specific meanings are unknown, more than one person must be aware that meanings are attached to visual images, whose form may be arbitrary with respect to their referents. In this light, how can a capacity that requires the idea of meaning, or specific meanings, to be shared by at least two individuals have arisen in a non-symbol-using society? Just as an individual within a modern human society is severely disadvantaged in social interaction if she/he cannot attribute meaning to visual image, so too would one be with this ability in an otherwise symbolically illiterate society. Believing that X means Y, or that X could mean A, B, or C, in such a social context would leave him/her in a minority of one—an effective definition of madness!

This suggests that although visual symbolism principally operates with respect to social interaction, we must look outside of the social sphere for the evolution of the capacity to create and read symbols. We need a situation in which an individual will have a selective advantage by virtue of possessing this capacity, even when it is absent in all conspecifics. A possible context is the use of "natural signs" in subsistence activity, epitomized by the use of animal tracks in hunting behavior. An ability to read meaning into such signs concerning animal behavior would increase foraging efficiency and give the unique symbol-reading (though not necessarily creating) individual a selective advantage. Since these signs exist whether they are being "read" or not, reference to them solves Davis's (1989:181) dilemma when he asks how the first image maker could know how to make an image if he had never seen one before—he/she had seen an image and attributed meaning to it, though it was not one created by a hominid.

If the use of natural signs provides the context for the development of the ability to attribute meaning, a second step is required for the full development of visual symbolism. The capacity to attribute meaning would need to be integrated with the three other abilities referred to above to allow the creation and use of symbols. I will draw on recent research in cognitive science to suggest that such integration arose from increased accessibility between mental modules possibly involving a hierarchization of cognitive processes. First we must discuss each of the four components of visual symbolism in turn. The first three—mark making, classification, and intentional communication—appear to have arisen early in human evolution as they are found in monkeys and apes.

MARK MAKING, CLASSIFICATION, AND COMMUNICATION

All terrestrial mammals create marks unintentionally in the course of their daily activities (e.g., footprints, tooth marks on plant material). All primates have the physical capacity to create marks intentionally. In laboratory contexts, apes have created striking paintings (Morris 1962) though, like language, this does not appear to occur in natural contexts. Artifacts marked by hominids are known from Lower Paleolithic contexts, such as cut marks on bones (e.g., Bunn 1981). The presence of manganese oxide and red ochre in many early Paleolithic contexts may imply intentional coloring activity (Wreschner 1980; Marshack 1981; Mellars 1989). Whether the patches of color created on cave walls, artifacts, or bodies had symbolic significance is unknown. These may have been similar to finger meanderings from many Paleolithic contexts which may constitute intentional, but "self-sufficient" marks, i.e. with no semantic content (Davis 1986).

Cheney and Seyfarth (1990) have assessed evidence for both classificatory skills and communicatory abilities among nonhuman primates in general, and vervet monkeys in particular. As Humphrey (1984) noted, the ability to recognize the likeness between things is essential for survival of any animal. Cheney and Seyfarth (1990) demonstrated that vervets can classify conspecifics in relation to abstract principles. For instance, membership of particular kinship classes can be recognized for animals of very different age and sex.

The vocal communications of vervets were also shown to convey meanings relatively independent of the context in which they are given. By an ingenious series of experiments using playback of predator alarm calls, Cheney and Seyfarth (1990; Seyfarth *et al.* 1980) showed that this involved reference to specific predators, rather than just various levels of emotional stimulation. Moreover, from their work and laboratory studies there is clear evidence for some degree of voluntary control over vocalizations. Inferring intentionality in communication is extremely problematic (Dennett 1988), but at least second-order intentionality (e.g., having beliefs and desire about beliefs and desires) appears to be present among vervet monkeys. Intentional communication, in the form of the teaching of tool use techniques, has been recognized among apes (Boesch 1993). In laboratory contexts and with the stimulus of a human instructor, apes have demonstrated far higher levels of intentional communication than in natural settings.

In summary, the ability to make marks, to classify, and to communicate intentionally appear to have arisen early in human evolution. These are shared by many nonhuman primates and can be safely assumed to be present among Pleistocene hominids. However, the fourth element of visual symbolism—the ability to attribute meaning to visual signs—appears absent among nonhuman primates in natural settings. Explaining the evolution of this ability is the first step in the two-stage evolutionary model for the origin of visual symbolism.

THE ATTRIBUTION OF MEANING

Apes and monkeys have very limited abilities at attributing mental states to other individuals (Cheney and Seyfarth 1990). They appear even more limited in their ability to attribute meaning to inanimate visual images. Apes have been shown to be capable of this ability when stimulated and instructed by a human in the context of acquiring sign language. Yet this ability has not been observed to arise without human stimulus in captive or natural contexts.

Cheney and Seyfarth (1990) described the inability of vervet monkeys to use inanimate secondary visual cues (ISVCs), which they describe as a central feature of nonsocial intelligence. ISVCs are visual images and objects which are not directly associated with their referent. They are epitomized by the tracks and trails of other species. As such they could be referred to as "indexes"—signs which do not visually resemble their referent but have a direct causal association (Chase 1991). Nonsocial intelligence is the ability to solve problems concerning inanimate objects or other species, and is contrasted with social intelligence—the ability to solve problems concerning interaction with conspecifics. For instance, pythons are a major predator on vervets yet the monkeys appeared unable to infer danger from a python track. Similarly, the presence of an antelope carcass in a tree, as if left by a leopard (another major predator), did not engender a sense of danger among the vervets.

This limited ability to use ISVCs contrasted with the vervet monkeys' ability to use primary visual cues and auditory cues. Two likely explanations were proposed by Cheney and Seyfarth. First, it may be the space/time displacement of an ISVC from its referent that makes their use particularly challenging for monkeys. Second, the lack of an ability to use ISVC may be because their intelligence is largely restricted to the social domain and within the latter, secondary visual cues play a very limited role. Vervet monkeys do not, for instance, intentionally create signs or follow each other's footprints. In contrast, auditory cues are frequently used in social interaction. As Cheney and Seyfarth state, "in their social interactions, the monkeys may never have needed to recognize that a visual cue can denote an absent referent" (1990:289).

Cheney and Seyfarth suggest that apes may be less constrained in their use of ISVCs than monkeys. Chimpanzees react to empty sleeping nests, and gorillas appear to sometimes deliberately follow the trails made by other gorillas when tracking a group. However, it appears likely that modern humans are the only primates to make extensive use of ISVCs, particularly those created unintentionally by other species (Hewes 1986, 1989; Davidson and Noble 1989). As such, this ability to attribute meaning to ISVCs must have arisen during the later stages of human evolution, i.e., after the divergence from apes at ca. 6 MYA.

ISVCS AND HUNTER-GATHERER FORAGING

Modern humans, and most notably those who live by hunting and gathering, are dependent on the use of such natural signs in their environment. These involve not only signs such as hoofprints but also marks left on vegetation and marks left on the ground by sleeping, rolling, and defecating. The amount of information that can be inferred from these by an experienced hunter is extensive. For instance, Lee (1979:212–213) described the use of tracks by a !Kung hunter. He explained how factors such as the size, shape, and depth of prints, the length of stride and gait, the amount of material kicked up between footfalls and that collected in the print since it was made may enable a forager to infer the type, age, sex, weight, state of health, and activity of the track maker, as well as the number of animals within a group and the time elapsed since the prints were made. Further descriptions of the use of tracks by modern foragers, illustrating their universal significance, are provided by Hill and Hawkes (1983:147), Estioko-Griffen (1986:38), Sullivan (1942:44, 67), Nelson (1983:46, 164, 173), and Jennes (1977:46).

ISVCs appear to share at least four of the five characteristics of human symbols described above:

1. They are often displaced in time and space from their referent. Hoofprints may have been made several days prior to being noticed.
2. The meaning of ISVCs are often shared, in the sense that there is a "correct" interpretation of the sign, and hunters, particularly when cooperating, will agree on this interpretation
3. The meaning of the ISVC will, however, vary with the knowledge of the observer. A hoofprint to a child may simply mean a deer. To an experienced hunter it may be an adult male deer that passed 24 hours previously.
4. ISVCs are often, if not always, unique. Their particular form depends on a complex range of factors such as ground and weather conditions. Moreover, hunters frequently correctly interpret new signs, most likely made up from a series of previously known primitives.

ISVCs may also be considered to a large extent arbitrary as to their referent. While the footprint of an animal may represent the shape of the hoof, and consequently be regarded as iconic, its referent for a hunter is not the animal's foot itself but the age, sex of the animal, the elapsed time since it passed, the speed it was going, and so forth. There is no visual resemblance between the mark and these referents. Similarly, other signs may be simply a scratch on a tree where deer rubbed their antlers, or nibbled shoots where deer had been feeding. These signs are abstract rather than representational. The major difference between human symbols and ISVCs is that the former are usually created intentionally as a means of communication, whereas natural signs are not. At least, they are not left as means of communication to humans; certain signs may serve as communication to conspecifics, such as to mark territory or attract the opposite sex. For humans, they are nonintentional symbols with which foraging efficiency can be increased once meaning has been attributed.

In light of the similarities between human symbols and ISVCs, it is likely that the same cognitive processes are used when recognizing and attributing meaning to the visual image. Consequently, the use of ISVCs may provide a selective context for the evolution of the ability to attribute meaning to a visual image. A forager competent at reading meaning into footprints will have an advantage in terms of locating and tracking game. There is now substantial evidence that the reproductive success of males in many hunter-gatherer groups is related to their foraging efficiency (Hill 1984; Kaplan and Hill 1985). Indeed, foraging efficiency is generally regarded as a route to reproductive success among many animals (Stephens and Krebs 1986). The large majority of the ethnographic descriptions of foragers using ISVCs relate to the hunting of large game. Hunters search for the tracks of specific species, search for tracks in general, use tracks and trails to pursue wounded animals, or use tracks when encountered in the course of other activities. It is unlikely that with prehistoric technology such hunting could have been conducted on a substantial scale if the ability to "read" ISVCs was absent, although the use of drive methods, rather than stalking, may have been possible. In this light the evidence for the origin of big game hunting is of considerable interest.

The appearance of big game hunting has been a major source of controversy during the last decade. Binford (1981, 1984, 1985) has argued that there is no convincing evidence for big game hunting prior to the early part of the Upper Pleistocene. The poor preservation of the early Paleolithic archaeological record, and the absence of methods to confidently distinguish between primary scavenging and hunting make Binford's assertions extremely difficult to evaluate, particularly for early Pleistocene sites from East Africa where the hunting/scavenging debate has focused (e.g., Isaac 1978, 1983; Binford 1981, 1985, 1986; Potts 1983, 1988; Bunn 1983; Bunn and Kroll 1986).

The data from the Middle Pleistocene of Europe are equally ambiguous. Villa (1990, 1991) suggests that Torralba and Aridos are simply too disturbed and impoverished to infer either hunting or scavenging strategies, as attempted by Binford (1987). Similarly, the faunal assemblages from Swanscombe and Hoxne are probably too small to either support Binford's (1985) assertion of scavenging, or to indicate an alternative (Stopp 1988). Gamble's (1987) "Man the Shoveler" model for carcass searching and thawing during the winters of the Middle Pleistocene is certainly attractive but as yet has no concrete test implications, although the off-site character of the archaeological record appears to offer some support. The interpretation of data from Klasies River Mouth has been as contentious as from any other site. Binford (1984) used patterning in skeletal part representation to argue that the Middle Stone Age people scavenged rather than hunted large bovids. Klein (1989) has pointed to problems with Binford's analysis, though also concluded that there was no regular hunting of large game animals prior to the Later Stone Age.

The data from the Middle Paleolithic of southwest France has been summarized by Chase (1986, 1989) and Mellars (1989) both of whom conclude, *contra* Binford, that some big game hunting was taking place. Several Middle Paleolithic sites are dominated by single species (e.g., Ilskaya, Tesik-Tash, Mauran, and levels within Combe Grenal) which has been taken to imply selective hunting rather than opportunistic scavenging. Mauran, in particular, has been frequently cited as a Middle Paleolithic site with evidence for specialized bison hunting in light of the large number of bison represented, though Straus (1990) notes that the Middle Paleolithic (rather than early Upper Paleolithic) date has yet to be verified. Chase (1986) argues that the skeletal remains from Combe Grenal are indicative of hunting rather than scavenging. Age profiles of horse from three of the Combe Grenal levels have catastrophic mortality profiles (Levine 1983) which have been used to infer a mass hunting strategy. Herd hunting is also likely at Mauran, La Cotte (Scott 1980).

The most reasonable conclusion to draw from these data and sets of conflicting interpretations is that some hunting of large game animals took place during the later part of the Middle Paleolithic, although this was on a relatively small scale, and involved relatively low degrees of planning depth as compared to the Upper Paleolithic. In this regard, it is likely that Middle Paleolithic hominids had the capacity to attribute meaning to ISVCs. And, as I have argued above, this capacity probably evolved because of the selective benefit in terms of increased foraging efficiency. The critical point is that the cognitive processes involved in such attribution of meaning to ISVCs during hunting are likely to be the same as those used by modern humans when attributing meaning to symbols intentionally created by other humans.

In summary, I have argued that at least by the Middle Paleolithic the four cognitive/physical processes involved in making and "reading" symbols were present—mark making, intentional communication, classification, and attribution of meaning. The first three of these arose early in human evolution, being shared by a range of nonhuman primates. The fourth appears distinctively human, and appeared relatively late in the context of subsistence activity. However, while it was later with respect to other cognitive processes, the four capacities appear to have been present for many thousands of years before they were integrated to form the "capacity for visual symbolism." This integration constitutes the second step of this evolutionary model for the origins of art. And to explore it we need to turn to the issue of mental modularity.

MODULARITY, ACCESSIBILITY, AND HIERARCHIZATION IN COGNITIVE EVOLUTION

My argument in this section is that these four components of visual symbolism became integrated during human evolution as a result of increased accessibility between the different mental modules, or cognitive domains, in which each of these abilities was originally located. This requires a preliminary discussion of the issue of mental modularity.

Mental Modularity

The distinction between social and nonsocial intelligence referred to above is just one characterization of a more general issue, namely, the extent to which the human mind is composed of a series of mental modules each dedicated to specific cognitive process or is a general-purpose information processor or learning mechanism. A trend within cognitive science has been to see it as the latter, although this has been expressed in several different ways, each providing an alternative perspective on the basic notion of mental modularity. While Fodor (1983) used the term *mental modules*, others have used *adaptive specializations* (Rozin 1976), *multiple intelligences* (Gardner 1983), *Darwinian algorithms* (Cosmides and Tooby 1987) and *cognitive domains* (Cheney and Seyfarth 1990). Each of these is an alternative interpretation/theory for the same basic idea that the human mind is composed of a series of discrete psychological mechanisms, which may be based in their own neurological structures. A further issue is the degree of accessibility of the flow of information and transference of psychological processes between mental modules.

The existence of some degrees of mental modularity is well established for many nonhuman animal species that display remarkable cognitive feats in some areas but are unable to apply such information processing to other tasks. For instance, bees are able to navigate over vast distances by using the sun, and salmon "remember" a specific river in which they spawned, yet neither of these show "intelligence" in other areas of their lives. In Cosmides and Tooby's terms, they have specific Darwinian algorithms for these tasks. Moreover, these cognitive abilities are to some extent innate, or preprogrammed, for they are too complex to be acquired by experience alone.

Chomsky (1972) used a similar argument to propose that human language is a discrete, partly innate, cognitive module. The speed at which children acquire language implies specialized psychological mechanisms for language acquisition, rather than the use of a general-purpose learning mechanism. After children have acquired complex grammatical structures and extensive vocabulary, they may still remain limited in other cognitive domains, such as the use and manipulation of numbers.

Fodor (1983) built on Chomsky's work to argue that not only language, but all processes of perception should be thought of as modules. He characterized these as computational elaborate, domain specific and informationally encapsulated. By the latter term he referred to the limited data base of knowledge they have access to. The most compelling example is the persistence of optical illusions when they are 'known' to be false: the visual perception module is encapsulated from such knowledge. Fodor argued that these perceptual modules typically work very quickly and without control (e.g., one cannot stop oneself hearing other than by physical means). He found support for such notions of modularity by drawing on evolutionary theory, and citing examples of cognitive pathologies which impair some mental processes but leave others intact.

Fodor made a contrast, however, between the modularity of perception and the generalized nature of cognition, or "central processes." According to Fodor, these are concerned with thought, reasoning, and problem solving—more generally with the fixation of belief. In central processes the information acquired by each module of perception is integrated to create a mental model of the world.

Multiple Intelligences

The "multiple intelligences" theory developed by Gardner (1983) makes no distinction between perception and cognition but characterizes the mind as composed of six types of intelligence:

1. Linguistic intelligence—that concerned with phonology, syntax, semantics, and pragmatics. Gardner noted that of all of the intelligences, that concerning language is the most well defined and is the least controversial as involving unique cognitive processes.
2. Musical intelligence—the ability of individuals to discern meaning and importance in sets of pitches rhythmically arranged and also to produce such metrically arranged pitch sequences as a means of communicating with individuals. Gardner argues that in spite of the similarity to linguistic intelligence in terms of the heavy reliance on auditory–oral abilities, musical abilities are mediated by separate parts of the nervous system and consist of separate sets of competences.
3. Logical mathematical intelligence—this concerns understanding the world of objects, actions, and the relationships between these. Gardner suggests that the work of Piaget, concerned with the development of thought through a series of distinct stages (sensorimotor to operational), was essentially concerned with logical mathematical intelligence alone. As such it is this intelligence that underlies math-

ematics and science, and is particularly developed and prized in Western societies.
4. Spatial intelligence—the capacity to perceive the visual world accurately, to perform transformations and modifications on one's initial perceptions and to be able to re-create aspects of one's visual experience even in the absence of relevant physical stimulation. The most elementary level of this is the ability to perceive the form of an object while more advanced levels concern the ability to perform mental rotations of three-dimensional objects.
5. Bodily-kinesthetic intelligence—the control of one's bodily motions and capacity to handle objects skillfully.
6. Personal intelligence—this has two aspects. On the one hand it concerns access to one's own thoughts and feelings, while on the other it concerns the ability to notice and make distinctions among other individuals, with particular relation to their moods, temperaments, motivations, and intentions.

Gardner argued that these intelligences interact with, and build on, each other but at the core of each is a computational capacity or information processing device which is unique to that particular intelligence and on which are based the more complex realizations and embodiments of that intelligence. He drew support from various sources such as studies of brain-damaged individuals, "idiot savants," prodigies, experimental psychology, and evolutionary theory to argue that the intelligences were indeed, to some extent, cognitively independent. Gardner laid considerable stress on the cultural environment and developmental process as to which types of intelligences will develop furthest. He particularly emphasized the stress laid in Western world on logical-mathematical intelligence and on this basis criticized Piagetian approaches for its almost sole focus on this when building a supposedly general theory of cognitive development.

Social and Nonsocial Intelligence

Gardner's notion of "personal intelligence" is essentially the same as that of "social intelligence" or "social knowledge." This is likely to have arisen early in human evolution (Humphrey 1976) and recent literature concerning nonhuman primates and cognitive development in children has given the notion of a distinctly "social intelligence" substantial support (e.g., Byrne and Whiten 1988; Cheney and Seyfarth 1990). Social intelligence is principally the ability to create and maintain with other group members a series of alliances that will be of benefit to oneself. This requires abilities to know and exploit the character of other individuals and their social relationships. Humphrey argues that for modern humans, social intelligence requires "being sensitive to other people's moods and passions, appreciative of their waywardness and stubbornness, capable of reading signs in their faces and equally the lack of signs, capable of guessing what each person's past holds hidden in the present for the future" (1984:4–5).

Cheney and Seyfarth (1990) contrasted social knowledge with nonsocial knowledge which concerns knowledge about other animal species, interactions between species, auditory and visual cues, i.e., both animate and inanimate objects, some of which may be relevant to survival. Nonsocial intelligence in this regard appears to combine elements of Gardner's spatial, logical-mathematical and bodily-kinesthetic intelligences. Cheney and Seyfarth conducted a unique series of experiments to evaluate the relative degrees of social and nonsocial intelligence in vervet monkeys. They found that, as theory predicted, vervet monkeys displayed much greater intelligence when interacting with conspecifics than with the nonsocial world. Many of the cognitive processes they used, such as the ability to classify conspecifics with regard to abstract categories, appeared inaccessible when interacting with either inanimate objects or members of other species. In particular, although the monkeys were very able to draw inferences from the behavior of other individuals, they were very poor at drawing inferences from secondary visual cues even when these appeared to be of considerable ecological value. For instance, the python is a major predator of the monkeys but they seemed unable to recognize the danger inherent in a recently made python track.

Darwinian Algorithms

A more extreme approach to mental modularity has been taken by Cosmides and Tooby (1987, 1989). They embed their approach more firmly in Darwinism arguing that the study of cognition makes the essential link between evolutionary theory and behavior. Rather than seeing domains or modules, they believe the mind is composed of a large series of psychological mechanisms each dedicated to a very specific problem, the solution of which was of benefit in the evolutionary environment of modern humans. Consequently, they stress the need to understand the character of that evolutionary environment since it provided the selective pressures for the particular psychological mechanisms that evolved and which we possess today. They coined the term *Darwinian algorithms* for those innate, domain-specific processes used for tasks such as kin recognition and foraging behavior. As such, Darwinian algorithms are now used in environments very different from those in which they evolved, and it is likely that many forms of modern behavior will show no adaptive relationship to the modern environment.

The relevance of these modularity perspectives on the human mind is that the four components of visual symbolism discussed above appear to be based in different mental modules. If we follow Gardner's categories, the making of marks is controlled by bodily-kinesthetic intelligence, classification of visual signs (based on discriminating between visual images) by spatial intelligence, intentional communication is based on linguistic intelligence (though perhaps also in personal intelligence), and the attribution of meaning by logical-mathematical intelligence. Alternatively, we might follow the simpler division between social and nonsocial intelligence and note that mark making and attribution of meaning concern physical objects and are based in nonsocial intelligence, whereas intentional communication and classification appear to have evolved as elements of social intelligence. If we were to adhere to an extreme modularity position, there would be no possibility for the integration of these to create cognitive modules such as the capacity for visual symbolism which would need to draw on several domains. Consequently, we must now deal with the issue of accessibility between mental modules.

Accessibility and Hierarchization

Accessibility refers to the degree of contact between mental modules. Rozin and Schull (1988) have argued that a critical feature of human intelligence is the high degree of accessibility, as compared to that of other primates. Much of our own experiences suggest that this is indeed the case. As Cheney and Seyfarth (1990) note, we use analogical reasoning not only to classify different types of kinship relations but in a diverse range of activities, such as when arguing about the taxonomic relations of hominid fossils. One of the major distinguishing features of modern human intelligence from that of other primates may be in our ability to extend knowledge gained in one context to new and different ones, and similarly to use psychological processes evolved to cope with one particular type of problem to an alternative, such as the application of those evolved for interaction with conspecifics to the interaction with other animal species. Gardner (1983) has recognized the significance of accessibility for human intelligence. While maintaining the idea that there are core psychological processes restricted to each intelligence, he notes that "in normal human intercourse, one typically encounters complexes of intelligences functioning together smoothly, even seamlessly in order to execute intricate human activities" (1983:279). The "central system" of Fodor (1983) is the ultimate in accessibility for here information from all perceptual modules is combined to create a model of the world.

As Rozin and Schull describe, the theory of accessibility suggests that the principal course of cognitive evolution and development has been from domain-specific cognitive processes to a more generalized intelligence. They suggest that this fits with the developmental phenomenon of *décalage*—the sequenced appearance of the same ability in different domains, and with aspects of development of number concepts and language. However, Greenfield (1991) has recently described the reverse process with reference to specific neural circuits. She described increasing modularity during development of those cognitive processes which control language and tool use. These become increasingly located in separate neural circuits.

Some resolution of this conflicting evidence may be found in the notion of hierarchization of cognitive processes during development and evolution. Gibson (1983, 1990, 1991a,b) has stressed the significance of hierarchical mental construction skills. By this she appears to mean the development of new mental structures, each constructed by an integration of those operating at a lower level in a cognitive hierarchy. Such development appears well established in child development. For instance Case (1985) describes how the transition from one cognitive stage to another involves the hierarchical integration of executive structures that were assembled during the previous stage, but whose form and function were considerably different. Gibson (1983; Parker and Gibson 1979) argue that in this respect human ontogeny does recapitulate its phylogeny, challenging Gould's (1976) arguments to the contrary. The process of hierarchization may be a principal means by which accessibility between mental modules occurs. It helps explain how modularity *and* accessibility may increase with experience during development.

CONCLUSION: ACCESSIBILITY AND THE MIDDLE/UPPER PALEOLITHIC TRANSITION

The appearance of visual symbolism is one of the major features of the Middle/Upper Paleolithic transition. This is now established as very mosaic and variable in character, but may nevertheless, as Mellars (1991) suggests, mark a major cognitive development in humankind. Along with Whallon (1989) and Binford (1989), he proposes that the dramatic behavioral changes—the appearance of art, bone, and antler technology, colonization of extreme environments, rapid culture change—reflect the appearance of fully modern language. I find this unlikely because of the considerable evidence for the evolution of brain structures relating to language much earlier in human evolution (Deacon 1989). But the notion of a cognitive threshold being passed is appealing and I would suggest that the transitions marks a dramatic increase in the degree of accessibility between mental modules or cognitive domains, i.e., the development of a modern, generalized human intelligence, though one still based in a modular cognitive architecture (Mithen 1992).

There is considerable evidence that the minds of Middle and Lower Paleolithic hominids exhibited higher degrees of modularity than those of modern humans. We can see in some aspects of their behavior, such as the production of stone tools, evidence for cognitive processes not dissimilar from those of modern humans (Gowlett 1984; Wynn 1989). Certainly the *chaînes opératoires* used in the Levallois technology of the Middle Paleolithic were no less complex than those of the Upper Palaeolithic (Boëda 1988). We can infer from such technology substantial planning depth, mental modeling, creative and flexible thought. Yet these cognitive processes appear absent in other domains of activity, notably interaction with the natural environment. The immense stability in the character of lithic and faunal assemblages during the Lower and Middle Paleolithic (Isaac 1977; Chase 1986; Binford 1989) reflects a form of interaction with the natural world very different from that of modern humans. This appears to lack the fine-grained adaptation to environmental variability characteristic of modern humans. This only occurs after the transition, as witnessed in the much higher degrees of culture change as technology is mapped onto changing environments. And this change is likely to derive from increased accessibility between mental modules. Those cognitive processes which had evolved in relation to social activity or stone working were transferred into other domains, hence creating a more generalized intelligence. One consequence of this, as discussed above, is the appearance of a hierarchy of cognitive processes with those at the higher levels drawing on and integrating those below, which had previously been encapsulated in separate cognitive domains.

The capacity for visual symbolism was probably just one of several new cognitive processes to arise. As I argued above, the four elements of the capacity for visual symbolism—mark making, classification, intentional communication, and attribution of meaning—were present prior to the Upper Paleolithic. But they were located in different cognitive domains within a modular mind. The higher-level cognitive processes to create and read symbols, requiring the integration of these separate elements, could only occur following the development of increased levels of accessibility between cognitive domains. With respect to the origin of art, this constituted the second of two critical steps, the first being the evolution of the ability to attribute meaning in the context of the use of animal tracks and trails in foraging activity.

REFERENCES

Bahn, P., 1986, No Sex, Please, We're Aurignacian's, *Rock Art Research* 3:99–120.
Bednarik, R.G., 1992, Palaeoart and Archaeological Myths, *Cambridge Archaeological Journal* 2:27–57.
Binford, L., 1981, *Bones: Ancient Men and Modern Myths*, Academic Press, New York.
Binford, L., 1984, *Faunal Remains from Klasies River Mouth*, Academic Press, New York.

Binford, L., 1985, Human Ancestors: Changing Views of Their Behavior, *Journal of Anthropological Archaeology* 4:292–327.
Binford, L., 1986, Comment on 'Systematic Butchery by Plio-Pleistocene Hominids at Olduvai Gorge, Tanzania' by H.T.Bunn and E.M.Kroll, *Current Anthropology* 27:431–452.
Binford, L.R., 1987, Were There Elephant Hunters at Torralba? in: *The Evolution of Human Hunting* (M.H. Nitecki and D.U. Nitecki, eds.), Plenum, New York, pp. 47–105.
Binford, L., 1989, Isolating the Transition to Cultural Adaptations: An Organizational Approach, in: *The Emergence of Modern Humans* (E.Trinkaus, ed.), Cambridge University Press, Cambridge, pp. 18–41.
Boëda, E., 1988, *Approche Technologique du Concept Levallois et Evaluation de son Camp d'Application: Etude de trois gisements Saliens etWetchseliens de la France Septentionale*, Unpublished Ph.D. dissertation, Université de Paris.
Boesch, C., 1993, Aspects of Transmission of Tool-use in Wild Chimpanzees, in: *Tools, Language and Cognition in Human Evolution* (K. Gibson and T. Ingold, eds.), Cambridge University Press, Cambridge, pp. 171–183.
Bunn, H.T., 1981, Archaeological Evidence for Meat Eating by Plio-Pleistocene Hominids from Koobi Fora, and Olduvai Gorge, *Nature* 291:574–577.
Bunn, H.T., 1983, Evidence on the Diet and Subsistence Patterns of Plio-Pleistocene Hominids at Koobi Fora, Kenya, and at Olduvai Gorge, Tanzania, in: *Animals and Archaeology: 1. Hunters and their Prey* (J. Clutton-Brock and C. Grigson, eds.), British Archaeological Reports International Series No. 163, Oxford, pp. 143–148.
Bunn, H.T., and Kroll, E.M., 1986, Systematic Butchery by Plio-Pleistocene Hominids at Olduvai Gorge, Tanzania, *Current Anthropology* 27:431–452.
Byrne, R.W., and Whiten, A. (eds.), 1988, *Machiavellian Intelligence: Social Expertise and the Evolution of Intellect in Monkeys, Apes and Humans*, Clarendon Press, Oxford.
Case, R., 1985, *Intellectual Development: Birth to Adulthood*, Basic Books, New York.
Chase, P., 1986, *The Hunters of Combe Grenal: Approaches to Middle Palaeolithic Subsistence in Europe*, British Archaeological Reports International Series No. 286, Oxford.
Chase, P., 1989, How Different Was Middle Palaeolithic Subsistence?: A Zoological Perspective on the Middle to Upper Palaeolithic Transition, in: *The Human Revolution: Behavioral and Biological Perspectives on the Origins of Modern Humans* (P. Mellars and C. Stringer, eds.), Edinburgh University Press, Edinburgh, pp. 321–337.
Chase., P., 1991, Symbols and Palaeolithic Artifacts: Style, Standardization and the Imposition of Arbitrary Form, *Journal of Anthropological Archaeology* 10:193–214.
Chase, P., and Dibble, H., 1987, Middle Palaeolithic Symbolism: A Review of Current Evidence and Interpretations, *Journal of Anthropological Archaeology* 6:263–293.
Chase, P., and Dibble, H., 1992, Scientific Archaeology and the Origins of Symbolism: A Reply to Bednarik, *Cambridge Archaeological Journal* 2:1, 43–51.
Cheney, D.L., and Seyfarth, R.M., 1990, *How Monkeys See the World*, University of Chicago Press, Chicago.
Chomsky, N., 1972, *Language and the Mind*, Harcourt, Brace, New York.
Cosmides, L., and Tooby, J., 1987, From Evolution to Behavior: Evolutionary Psychology as the Missing Link, in: *The Latest on the Best: Essays on Evolution and Optimality* (J.Dupre, ed.), MIT Press, Cambridge, MA, pp. 276–306.
Cosmides, L., and Tooby, J., 1989, Evolutionary Psychology and the Generation of Culture, Part I, *Ethology and Sociobiology* 10:29–49.
Davidson, I., 1990, Bilzingsleben and Early Marking, *Rock Art Research* 7:52–56.
Davidson, I., 1992, There's No Art 'To Find the Mind's Construction' In Offence: Reply to R. Bednarik 1992, 'Palaeoart and Archaeological Myths', *Cambridge Archaeological Journal* 2:52–57.

Davidson, I., and Noble, W., 1989, The Archaeology of Perception: Traces of Depiction and Language, *Current Anthropology* 30:125–155.

Davis, W., 1986, The Origins of Image Making, *Current Anthropology* 27:193–215.

Davis, W., 1989, Finding Symbols in History, in: *Animals into Art* (H. Morphy, ed.), Unwin Hyman, London, pp. 179–189.

Deacon, T., 1989, The Neural Circuitry Underlying Primate Calls and Human Language, *Human Evolution* 4:367–401.

Delluc, B., and Delluc, G., 1978, Les manifestations graphiques aurignaciennes sur support rocheux des environs des Eyzies (Dordogne), *Gallia Prehistoire* 21:213–438.

Dennett, D., 1988, The Intentional Stance in Theory and Practice, in: *Machiavellian Intelligence: Social Expertise and the Evolution of Intellect in Monkeys, Apes and Humans* (R.W. Byrne and A. Whiten, eds.), Clarendon Press, Oxford, pp. 180–202.

Duff, I.A., Clark, G.A., and Chadderdon, T.J., 1992, Symbolism in the Early Palaeolithic: A Conceptual Odyssey, *Cambridge Archaeological Journal* 2(2):211–229.

Estioko-Griffen, A., 1986, Daughters of the Forest, *Natural History* 95(5):36–42.

Fodor, J., 1983, *The Modularity of Mind*, MIT Press, Cambridge, MA.

Gamble, C., 1987, Man the Shoveler: Alternative Models for Middle Pleistocene Colonization and Occupation in Northern Latitudes, in: *The Pleistocene Old World* (O. Soffer, ed.), Plenum Press, New York, pp. 81–98.

Gardner, H., 1983, *Frames of Mind: The Theory of Multiple Intelligences*, Basic Books, New York.

Gibson, K., 1983, Comparative Neurobehavioral Ontogeny and the Constructionist Approach to the Evolution of the Brain, Object Manipulation and Language, in: *Glossogenetics: The Origin and Evolution of Language* (E. de Grolier, ed.), Harwood Academic, New York, pp. 52–82.

Gibson, K., 1990, New Perspectives on Instincts and Intelligence: Brain Size and the Emergence of Hierarchical Mental Construction Skills, in: *"Language" and Intelligence in Monkeys and Apes* (S.T. Parker and K. Gibson, eds.), Cambridge University Press, Cambridge, pp. 97–128.

Gibson, K., 1991a, Continuity Versus Discontinuity Theories of the Evolution of Human and Animal Minds: Comment on 'Language, Tools and Brain: The Ontogeny and Phylogeny of Hierarchically Organized Sequential Behavior, by P.M. Greenfield, *Behavioral and Brain Sciences* 14:560–561.

Gibson, K., 1991b, Tools, Language and Intelligence: Evolutionary Implications, *Man* 26:255–264.

Gould, S.J., 1976, *Ontogeny and Phylogeny*, Harvard University Press, Cambridge, MA.

Gowlett, J., 1984, The Mental Abilities of Early Man: A Look at Some Hard Evidence, in: *Human Evolution and Community Ecology* (R. Foley, ed.), Academic Press, New York, pp. 167–192.

Greenfield, P.M., 1991, Language, Tools and Brain: The Ontogeny and Phylogeny of Hierarchically Organized Sequential Behavior, *Behavioral and Brain Sciences* 14:531–595.

Hahn, J., 1972, Aurignacian Signs, Pendants, and Rare Objects in Central and Eastern Europe, *World Archaeology* 3:252–266.

Hahn, J., 1984, Recherches sur l'art paléolithique depuis 1976, in: *Aurignacien et Gravettian en Europe*, Vol. 3 (J.K. Kozlowski and R. Desbrosses, eds.), Etudes et Recherches Archéologiques de l'Université de Liège No. 13, pp. 79–82.

Haskovec, I.P., and Sullivan, H., 1989, Reflections and Rejections of an Aboriginal Artist, in: *Animals into Art* (H. Morphy, ed.), Unwin Hyman, London, pp. 57–74.

Hewes, G., 1986, Comment on 'The Origins of Image Making' by W. Davis, *Current Anthropology* 27:193–215.

Hewes, G., 1989, Comment on 'The Archaeology of Perception: Traces of Depiction and Language' by I. Davidson and W. Noble, *Current Anthropology* 30:145–146.

Hill, J., 1984, Prestige and Reproductive Success in Man, *Ethology and Sociobiology* 5:77–95.

Hill, K., and Hawkes, K., 1983, Neotropical Hunting Among the Ache of Eastern Paraguay, in: *Adaptive Responses of Native American Indians* (R. Hames and W. Vickers, eds.), Academic Press, New York, pp. 139–188.
Humphrey, N., 1976, The Social Function of Intellect, in: *Growing Points in Ethology* (P.P.G. Bateson and R.A. Hinde, eds.), Cambridge University Press, Cambridge, pp. 303–317.
Humphrey, N., 1984, *Consciousness Regained*, Oxford University Press, Oxford.
Isaac, G.I., 1977, *Olorgesaille, Archaeological Studies of a Middle Pleistocene Lake Basin in Kenya*, University of Chicago Press, Chicago.
Isaac, G.I., 1978, The Food-sharing Behavior of Proto-human Hominids, *Scientific American* 238:90–108.
Isaac, G.I., 1983, Bones in Contention: Competing Explanations for the Juxtaposition of Early Pleistocene Artifacts and Faunal Remains, in: *Animals and Archaeology: 1. Hunters and Their Prey* (J. Clutton-Brock and C. Grigson, eds.), British Archaeological Reports International Series No. 163, Oxford, pp. 3–19.
Jennes, D., 1977, *The Indians of Canada*, 7th ed., University of Toronto Press, Ottawa.
Kaplan, H., and Hill, K., 1985, Hunting Ability and Reproductive Success among Male Ache Foragers: Preliminary Results, *Current Anthropology* 26:131–133.
Klein, R.G., 1989, Biological and Behavioral Perspectives on Modern Human Origins in Southern Africa, in: *The Human Revolution* (P. Mellars and C. Stringer, eds.), Edinburgh University Press, Edinburgh, pp. 529–546.
Layton, R., 1986, Comment on 'The Origins of Image Making' by W. Davis, *Current Anthropology* 27:205.
Lee, R.B., 1979, *The !Kung San: Men, Women and Work in a Foraging Society*, Cambridge University Press, Cambridge.
Levine, M., 1983, Mortality Models and the Interpretation of Horse Population Structure, in: *Hunter-Gatherer Economy in Prehistory* (G.N. Bailey, ed.), Cambridge University Press, Cambridge, pp. 23–46.
Marshack, A., 1981, On Palaeolithic Ochre and the Early Use of Color and Symbol, *Current Anthropology* 22:181–191.
Marshack, A., 1990, Early Hominid Symbol and the Evolution of Human Capacity, in: *The Emergence of Modern Humans* (P. Mellars, ed.), Edinburgh University Press, Edinburgh, pp. 457–498.
Mellars, P., 1989, Major Issues in the Emergence of Modern Humans, *Current Anthropology* 30:349–385.
Mellars, P., 1991, Cognitive Changes and the Emergence of Modern Humans, *Cambridge Archaeological Journal* 1:63–76.
Mithen, S.J., 1992, From Domain Specific to Generalized Intelligence: A Cognitive Interpretation of the Middle/Upper Palaeolithic Transition, in: *The Ancient Mind* (C. Renfrew and E. Zubrow, eds.), Cambridge University Press, Cambridge, pp. 29–39.
Morphy, H., 1989, On Representing Ancestral Beings, in: *Animals into Art* (H. Morphy, ed.), Unwin Hyman, London, pp. 144–160.
Morris, D., 1962, *The Biology of Art*, Methuen, London.
Nelson, R.K., 1983, *Make Prayers to the Raven: A Koyukon View of the Northern Forest*, University of Chicago Press, Chicago.
Parker, S.T., and Gibson, K., 1979, A Model of the Evolution of Language and Intelligence in Early Hominids, *Behavioral and Brain Sciences* 2:367–407.
Potts, R., 1983, Foraging for Faunal Resources by Early Hominids at Olduvai Gorge, Tanzania, in: *Animals and Archaeology: 1. Hunters and Their Prey* (J. Clutton-Brock and C. Grigson, eds.), British Archaeological Reports International Series No. 163, Oxford, pp. 51–62.
Potts, R., 1988, *Early Hominid Activities at Olduvai*, Aldine de Gruyter, New York.

Rozin, P., 1976, The Evolution of Intelligence and Access to the Cognitive Unconscious, in: *Progress in Psychology*, Vol. 6 (J.N. Sprague and A.N. Epstein, eds.), Academic Press, New York, pp. 245–277.

Rozin, P., and Schull, J., 1988, The Adaptive-evolutionary Point of View in Experimental Psychology, in: *Stevens' Handbook of Experimental Psychology*, Vol. 1 (R.C. Atkinson, R.J. Herrnstein, G. Lindzey, and R.D. Luce, eds.), Wiley, New York, pp. 503–546.

Scott, K., 1980, Two Hunting Episodes of Middle Palaeolithic Age at La Cotte de Saint-Brelade, Jersey (Channel Islands), *World Archaeology* 12:137–152.

Seyfarth, R.M., Cheney, D.L., and Marler, P., 1980, Vervet Monkey Alarm Calls: Semantic Communication in a Free Ranging Primate, *Animal Behavior* 28:1070–1094.

Stephens, D.W., and Krebs, J.R., 1986, *Foraging Theory*, Princeton University Press, Princeton.

Stopp, M., 1988, A Taphonomic Analysis of the Hoxne Site Faunal Assemblages, Unpublished M.Phil thesis, University of Cambridge.

Straus, L.G., 1990, On the Emergence of Modern Humans, *Current Anthropology* 31:63–65.

Sullivan, R.J., 1942, *The Ten'a Food Quest*, The Catholic University of America Press, Washington, DC.

Taylor, L., 1989, Seeing the 'Inside': Kunwinjku Paintings and the Symbol of the Divided Body, in: *Animals into Art* (H. Morphy, ed.), Unwin Hyman, London, pp. 371–389.

Villa, P., 1990, Torralba and Aridos: Elephant Exploitation in Middle Pleistocene Spain, *Journal of Human Evolution* 19:299–309.

Villa, P., 1991, Middle Pleistocene Prehistory in Southwestern Europe: The State of Our Knowledge and Ignorance, *Journal of Anthropological Research* 47:193–217.

Whallon, R., 1989, Elements of Cultural Change in the Later Palaeolithic, in: *The Human Revolution: Behavioral and Biological Perspectives on the Origins of Modern Humans* (P. Mellars and C. Stringer, eds.), Edinburgh University Press, Edinburgh, pp. 433–454.

White, R., 1989, Production Complexity and Standardization in Early Aurignacian Bead and Pendant Manufacture: Evolutionary Implications, in: *The Human Revolution: Behavioral and Biological Perspectives on the Origins of Modern Humans* (P. Mellars and C. Stringer, eds.), Edinburgh University Press, Edinburgh, pp. 366–390.

Wobst, H.M., 1977, Stylistic Behavior and Information Exchange, in: *Papers for the Director: Research Essays in Honor of J.B.Griffen* (C.E. Cleland, ed.), Anthropological Papers No. 61, Museum of Anthropology, University of Michigan, Ann Arbor, pp. 317–342.

Wreschner, E., 1980, Red Ochre and Human Evolution: A Case for Discussion, *Current Anthropology* 21:631–644.

Wynn, T., 1989, *The Evolution of Spatial Competence*, University of Illinois Press, Urbana.

Part V
OVERVIEW

Chapter 12

The State of Evolutionary Archaeology
Evolutionary Correctness, or the Search for the Common Ground

ROBERT L. BETTINGER AND PETER J. RICHERSON

The problem, straight off, is that the common ground is not as clear as it should be. To the reader it must seem that evolutionary archaeologists are individually certain they are standing on the common ground, equally certain that their colleagues are not, and that the job is mainly to bring the undecided and misguided into the fold. Competition brings out the best in some products, possibly, but perhaps not in theory. All of the authors here would surely agree that the evolutionary processes that operate in the human case are complex and still poorly understood. After all, the biological disciplines from which much of the basic theory is being borrowed have historically been, and continue to be, filled with contentious debate over the processes of organic evolution. In the case of humans, we deal mainly with one species, but also with the relatively unstudied, and decidedly controversial, complexities of cultural evolution. Under these circumstances, competition between theorists plays a useful role only if the reader is not misled by the rhetorical excesses of self-advertisement and re-

---

ROBERT L. BETTINGER • Department of Anthropology, University of California, Davis, California 95616. **PETER J. RICHERSON** • Division of Environmental Studies, University of California, Davis, California 95616.

mains diligent in discriminating between clear thinking and problem solving, on the one hand, and dogma and the strawman, on the other. It helps to recognize that at the early stages of the exploration of a new field, even the main lines of fruitful inquiry remain shrouded in ignorance. Polemical essays are best taken merely as claims that a particular line of inquiry is promising enough to be worth pursuing. Read literally, some of the arguments here verge on claiming a unique theoretical correctness for certain evolutionary concepts, processes, or lines of empirical inquiry. Many such claims are highly abstract, which should trigger a warning signal in the reader's mind. Scientific issues are generally settled with concrete tests of cogency (often involving mathematical tools) or critical data (collected with effort and analyzed with care). In the case of the complex and diverse subject matter of evolution, many models, much data, and considerable time are invariably required to arrive at definitive answers. It is hard to think of a case where important issues in evolutionary biology were settled by abstract, *a priori* claims.

Things are not yet badly out of hand in evolutionary archaeology, to be sure. Nevertheless, we want to use this commentary to reaffirm the common ground shared by all evolutionary approaches and the diversity of subject matters that can be legitimately pursued under the broader neo-Darwinian umbrella. In particular, we stress the pluralism and productivity of Darwinian methods as an antidote to excesses inadvertently produced by sincere advocacy of particular points of view. Surely one reason for the century-long boom of evolutionary inquiry touched off by Darwin and Wallace is precisely that Darwinian theory accommodates a virtually limitless range of interesting, fruitful projects. It is perhaps this diversity that causes us sometimes to forget the basic commonalities that unite the Darwinian enterprise. This is unfortunate because scientific disciplines are cooperative ventures in which one needs all of the friends one can get. It is worth keeping in mind here that the currently strong anti-Darwinian sentiment in the social sciences derives from individuals who manage to set aside their otherwise substantial intellectual differences when opposing evolutionary interpretations. Perhaps we should follow the model.

We wish to make it clear, in any case, that we have no fundamental quarrel with any of the arguments presented in this volume. Most of the individual theoretical expositions, including our own, pertain to limited subject matters, the differences and contradictions between which are more apparent than real. In this essay we make several points that we hope almost every evolutionary archaeologist would agree are virtually motherhood claims. We aim to be trite and claim no originality for the following points. For understandable reasons, we neglect our fundamental commonalities in the heat of debate, but we ought sometimes to remember them.

PARADIGMS REQUIRE PARADIGMATIC CASE STUDIES

The volume contributors are painfully aware of the limited number of actual applications of evolutionary theory to archaeological data. Scientific revolutions, of course, are frequently started and largely decided without critical applications. Copernican astronomy, for instance, replaced Ptolemaic astronomy well before the Copernican system could produce superior predictions of planetary movements (Kuhn 1962:68, 75–76, 153). Einstein's theory of relativity was widely accepted long before anyone was clever enough to instrument a critical test (Kuhn 1962:26, 154). In both instances, the scientific fields involved simply adopted a theoretical structure on grounds other than proven ability to predict or explain data—mostly on faith (eventually sustained) that the view being advanced was so compelling it somehow had to be correct. Sooner or later, however, we are going to have to do better. Completed paradigms must provide a battery of research designs that articulate theory with methodology, instrumentation, and data.

Evolutionary archaeology, which surely has the requisite general theory and instrumentation, still lacks proven research routines that show how one might reasonably address real data within this larger conceptual structure. We have in mind here what might be called *paradigmatic case studies*, or *exemplars* as Geier calls them (1988:35): closely specified research designs, the principles of which, if carefully followed, can be successfully applied almost endlessly. In evolutionary biology, one thinks of the staple studies of gene frequency change under artificial selection, studies of selection (or lack thereof) in natural populations, and the dissection of adaptive significance of bits of physiology, anatomy, and behavior. In archaeology, one thinks of the classic case studies in "ceramic sociology," mortuary analysis, and probabilistic regional survey that helped convert the New Archaeology from a largely theoretical endeavor into a working paradigm. Evolutionary culture theory needs a few of these. Students need to be pointed to subject matters (data) that can fairly reliably be depended on to yield evolutionarily interesting information. This is clearly why, within evolutionary archaeology, evolutionary culture theory has lagged behind evolutionary ecology and Darwinian psychology, both of which have converted evolutionary principles into successful working methodologies. This leads us to our second point.

EVOLUTIONARY ECOLOGY IS EVOLUTIONARY RESEARCH

Two of the three areas that dominate contemporary evolutionary anthropological inquiry—evolutionary culture theory and Darwinian psychology, are well represented in this volume. Evolutionary ecology, the third, most active—

and arguably most successful—area of evolutionary research, is not. This is surprising because evolutionary ecology embraces a well-developed research program that is heavily given to the testing of a relatively limited range of fairly specific hypotheses about mating and subsistence. Despite this, a large contingent of evolutionary archaeologists, including some contributors to this volume, find evolutionary ecology, and studies guided by models of optimization particularly, devoid of evolutionary content.

In this volume, the logic behind this criticism can be found in the Abbott, Leonard, and Jones analysis of the Parry–Kelly hypothesis that the North American transition from biface to flake lithic technology was a functional response to growing regional sedentism. The analogy applies because, closely inspected, the Parry–Kelly account recapitulates the key assumptions of evolutionary ecology, which attempts to demonstrate how specific behaviors (e.g., subsistence) are the result of economizing measures contributing to fitness. In the Parry–Kelly case the use of time and lithic resources is said to change as an efficient functional response to settlement change.

Abbott *et al.* find this unsatisfactory because the shift to sedentism that entrains the accommodative change in lithic technology is itself not explained with reference to evolutionary forces. The argument, then, is functional, not evolutionary, and, by implication, unsatisfactory. This argument seems to imply that if we cannot solve all evolutionary problems simultaneously, we cannot solve any of them. If that's true, we're sunk. The vast everything-is-connected-to-everything nature of the real ecological and evolutionary world means that success depends on breaking problems down into manageable chunks. The bulk of evolutionary, ecological, and physiological research and explanation proceeds this way. Thus, the question posed to a physiologist, "Why is this dog panting?" is more appropriately and directly answered by saying "To regulate its body temperature," than by a protracted explanation involving the evolution and natural history of dogs and warm-bloodedness. In responding without direct reference to evolutionary processes, the physiologist does not question that this panting is the result of a long evolutionary history. This history, however, is beside the point directly at hand ("Why is this dog panting?"). Further, the answer ("To regulate its body temperature") is surely useful, despite its lack of direct reference to evolutionary history. The evolutionary connection is potentially there because evolutionary theory is a fertile source of hypotheses in regard to the functional design of organs and organisms and because functional responses frequently contain important clues about evolutionary history that are worth paying attention to.

Darwin's beautiful "It is interesting to contemplate a tangled bank" final paragraph of the *Origin* is an invitation to the mystically inclined, and we encourage anyone so inclined to travel down that path (Darwin 1859). "There is grandeur in this view of life, with its several powers, having been originally breathed by the Creator into a few forms or one, and that, whilst this planet has gone on cycling according to the fixed law of gravity, from so simple a beginning endless forms most beautiful and most wonderful have been, and are being, evolved."

Our basic point is that the broad view of evolutionary research is to be preferred to any narrow view, so long as the scope of our *scientific* vision causes us to do more than merely contemplate with bemusement the grandeur of the tangled bank. There are so many potentially fruitful levels of evolutionary inquiry, all of them interesting. General, large-scale processes such as the spread of maize agriculture are important, to be sure. The small-scale functional processes, however, are no less interesting, if only because the process of selection operates fundamentally at the most intimate scales of space and time, and gains its power only by accumulating a multitude of tiny events to produce the grand tapestry. It is critically important that we be mindful that these different levels exist and that we appropriately match data, question, and process. We needn't (and shouldn't) invoke large-scale process to account for data more readily accounted for by functional hypotheses generated on the basis of clues provided by adaptive design, which are just too often prematurely ignored as "adaptive just-so stories"—which is our next point.

MODELS *ARE* JUST-SO STORIES

It is critical to keep in mind the essential distinction between our models and the subject matters we want them to capture. All of us join Gould and Lewontin, in decrying adaptive "just-so" stories. What we need to remember is that the "just-so" part is not what makes them objectionable. They are objectionable when, as frequently happens, they are merely asserted as true rather than set forth as evolutionary hypotheses with potentially rich and testable implications. *Model*, after all, is just a fancy term for "just-so story." Scientific advance requires that we think about our models just this way. Pursuit of the truth via the scientific method can only occur if it is assumed one doesn't have to discover the real "God's truth" but instead crude approximations and modest improvement in older, even cruder approximations. Just-so stories, optimality arguments, or whatever, constitute the necessary stepwise constructions. Their unsatisfactoriness relative to some Olympian standard of explanation is a necessary cost of doing business. The job is to follow simple rules provided by theory in generating a just-so story that matches the data more closely than the just-so stories generated in other ways, and more closely than the last try.

The optimal just-so stories of evolutionary ecology commonly imagine what subsistence behavior in a given context would look like if individuals "acted-as-if" they were optimizing momentary energetic rate of return. If these expectations are in closer accord with observed data than expectations arising from other "just-so, acted-as-if" accounts, we have gained a glimpse of the processes shaping human behavior. Robert Brandon (1990) discusses the contrast between "how actually" and "how possibly" explanations of evolutionary phenomena. The criteria for producing plausible "how actually" accounts of the evolution of a particular structure or behavior are dauntingly difficult, indeed impossible, according to Brandon, when historical contingencies intrude. "How possibly" explanations are less explicitly specified and supported, yet far from devoid of content, and there are clear criteria for choosing among alternative formulations. It will help dialogue to understand that, given time's ravages, few archaeologists will ever be privileged to participate in constructing a "how actually" explanation.

In our case, the problem of Darwin's tangled bank is serious. Human evolutionary processes are so complexly intertwined at many levels that even the simplest problems frequently resist definitive solution. In the face of such complexity, the theorist turns from analytical models to simulation, and the empiricist to the field or laboratory for more data. Simulation is a powerful tool, but using it moves us into the murky realm where the model itself is merely as tangled as the real world it mimics. If one explains a case in terms of correspondences between a simulation result and the real world when the model itself is too complex to understand, it is not clear than any real progress has been made. As empirical data pile up, it often seems that every proffered explanation can be rejected as incomplete if not downright wrong. What Darwin and Brandon are telling us is that a complete understanding of most evolutionary events is beyond the reach of current methods. Such is the typical complexity of ecology and evolution that we perhaps can never hope to get beyond "how possibly" explanations, except in a handful of special cases. Sensible scientists conduct their investigations with these limits in mind, and know when to turn the case over to the mystics. Just-so stories become egregious when such limits are ignored.

INTENT IS AN APPROPRIATE SUBJECT OF STUDY AND COMPONENT OF THEORY

The problem of intent arises mainly in teleological arguments that conflate intent with outcome. The principle is that intent (i.e., motivation) and behavioral outcome are not the same thing, and that one should not assume observed behaviors are the direct results of the motives that seem most likely to produce those behaviors. This is all the more true because one frequently has access to behavioral information only at the population level, whereas motives and intent always reside at the individual level. In light of this, it is important to keep in mind that, because the actions of individuals have population-level consequences that are difficult to intuit, group consequences do not follow in any simple way from individual intent.

This dictum is easy to forget in the case of humans because of the Lamarckian nature of cultural inheritance. Individual intentions—decisions at any rate—figure as forces in most theories of cultural evolution as is frequently noted in this book. It is hard to see how one can build a fire wall that separates intentional and nonintentional processes; they seem to interact intimately. In any case, it will not do simply to ignore the effects of individual decision-making in the case of cultural evolution. (Indeed, such is not really possible even in the absence of culture, as the effect of mate choice in sexual selection shows.) The idea that human agents simply choose their culture is, of course, an equally misleading metaphor. Group consequences cannot be reduced to individual behavior any more than the behavior of individuals can be reduced to the behavior of groups. We should be wary of substituting folk psychology for real psychology in theorizing about intent, as evolutionary psychologists constantly warn. It would not seem to follow that, however, because some theories of individual intent are grossly teleological that all of them are or that intent is not a legitimate matter of study and theorizing. It is quite clear, for instance, that all sorts of organisms are capable of learning, which requires intent, by which we mean an abstract goal against which outcomes of various behaviors are evaluated. It is, moreover, specifically with reference to models involving individual intent that evolutionary theory has consistently proven itself superior to traditional anthropological theory.

Among the most important contributions of evolutionary theory to anthropology is the observation that, because individuals are always important in evolutionary processes, it is safest to initiate an evolutionary inquiry under the assumption that individuals will be motivated to do what they do principally out of self-interest and will assume that others are operating similarly. This means we should always be skeptical of arguments in which individuals are said to be motivated in ways that cause them to do things potentially not in their interest. Thus, the collective argument, that individuals engage in cooperative behaviors to benefit the groups to which they belong, is immediately vulnerable to a counterargument in which one asks, given a population of such cooperators, "What would happen to an individual who reaped the benefits of cooperation without actually cooperating?" The answer, of course, is the crux of the prisoner's dilemma and other public goods scenarios, which are primarily concerned with intent and distrust.

In anthropology, it is surely this replacement of groups/cultures by self-interested individuals as basic units of analysis that is the most revolutionary consequence of evolutionary theory. Once the trick is learned, it is child's play to dismantle almost any traditional anthropological interpretation in which the actions of individuals are explained with reference to group-level needs (e.g.,

Bettinger 1991). That this is so easily done has been widely generalized as indicating that groups/cultures have no special evolutionary properties worth attending to and one frequently encounters the argument that group selection explanations are fatally flawed *a priori*. The problem with this, as cultural anthropologists are quick to point out, is that, however well-warranted in theory, the assertion is in obvious conflict with the empirical record, where evidence of human cooperation and individual self-sacrifice is everywhere. Human behavior, in short, does seem to relate to groups and cooperative behaviors that are rare in the animal world. This is our next point.

THE EMERGENT PROPERTIES OF GROUPS SHOULD NOT BE IGNORED

It must not be forgotten that the population level is an irreducible component of any Darwinian theory. Ernst Mayr (1982) described Darwin's revolution as "population thinking." In the simplest cases, the population is just the evolving pool of genes or ideas that individuals draw on each generation. But more complex relationships between the individual and population levels are commonly studied in population biology. For example, in the cases of density- or frequency-dependent selection, the statistical properties of the population are arguments in the fitness function bearing on individual variation. Just as there is no fire wall preventing individual-level intention from figuring as an evolutionary force affecting a population, there is none preventing the population state from feeding back downward to affect the individual level.

The human case is, on the basis of *prima facie* empiricism, an extreme case of effects feeding down to individuals from the population level. Human groups exhibit cooperation, coordination, and division of labor on a scale that is reminiscent of the social insects. Even in the case of hunter-gatherers, the ethnolinguistic unit is typically organized by marriage systems, linguistic and other symbolic markers, ethical norms specifying different treatment of ingroup and outgroup members, and the like. The human group is salient for individuals in a way that is unique. In the social insects, the cooperating unit is a kin group, as is usually the case for any sustained cooperative unit among other animals. In humans it is the whole breeding population—or at any rate large fractions of it—that is organized. Humans are quite prepared to cooperate with quite distantly related individuals provided they are members of some culturally defined ingroup.

Evolutionary social scientists have carried over from biology, and perhaps even exaggerated, the dogma that group selection never works. One should, of course, be leery of arguments in which the presence of unified groups is uncritically assumed, yet there is nothing in evolutionary theory that absolutely precludes the action of group selection. Darwin, William Hamilton, and Richard Alexander—whose evolutionary credentials and commitment to evolutionary general theory are presumably beyond dispute—have all been prepared to see a role for group selection in the special case of humans. The evolutionary origin of ultra-sociality in humans is one of the key puzzles for evolutionary social science and it would be perverse in the extreme to reject summarily all group selection explanations of such phenomena merely out of loyalty to a general theoretical argument, no matter how well justified that argument is. Every species is a special case in some way or another, and we humans might just be unique in the processes surrounding the evolution of our obviously peculiar social organization. This brings us to the matter of the use of analogies between the evolutionary processes governing human and nonhuman systems.

ANALOGUES BETWEEN EVOLUTIONARY PROCESS IN NATURAL AND CULTURAL SYSTEMS ARE JUST ANALOGUES

There is much to be learned by asking in what way human systems—cultural systems in particular—are processually akin to natural ones. This, in fact, is the inspiration that ultimately stands behind neo-Darwinian inquiries in anthropology. All the same, we need to keep very clear in our own minds that culture is not exactly the same as any other natural system. The units of culture (should they exist) are not genes, neither are they literal viruses. The question is in what essential properties are culture processes operationally identical, or nearly so, to the evolutionary processes to which genes or viruses are subject. One way to answer this is by starting at the other end and asking in what way the two systems are different and under what circumstances would these differences be likely to reveal themselves. For instance, the heredity of both genes and viruses is strongly conditioned by the fact that they have a material existence as nucleic acids, which is not true of culture. Transmission by social learning is something *sui generis*. It is quite sensible to use analogies for inspiration and to borrow them as tools, but it is not so useful to engage in extended debate about just which analogies are the most inspirational.

DO THE MATH

One way to resolve whether one needs the theoretical baggage that goes with any of the analogies one is tempted to draw between evolutionary processes in the cultural and natural world is to ask whether the analogy requires the use of special quantitative algorithms designed specifically for use with that natural system. One wants to ask, for example, whether thinking about culture as viruses requires math any different from that which is already in use. That culture has viruslike properties, for instance, would surely not surprise Cavalli-Sforza and Feldman (1981), who opened their book on culture transmission theory with an epidemiological model and introduced the vertical/oblique/horizontal transmission terminology from the same field. Deep down, one suspects, most of the glosses arising in reference to analogies between natural and cultural systems have little effect on the structure of the models we'd use. Equations designed to capture selective processes acting on individuals, for example, often work just as well for groups. This is not to say either that natural–cultural similarities don't exist or that natural–cultural differences aren't important. It's just to say that one way to get past the ambiguities inherent in verbal analogies is to put our models down as simply and concretely as we can—quantitatively if possible. If there is more inspiration to be dragged out of biological analogies, very good. Just write down a model and show us what we've missed.

The debate over the term *cultural selection* is a case in point. A good many evolutionary archaeologists are surprisingly sensitive to the use of the term on the grounds that it implies a separation between cultural evolutionary processes and natural ones. In a sense, of course, dependence on socially transmitted information makes cultural systems unique, or at least different, from most natural systems where genetic information is more important. The difference between cultural and most natural systems seems at least as great, if not greater than the differences between sexual and asexual systems of reproduction. Assertions of this sort, however, are not what we are talking about. *Cultural selection* is a reasonable term if one can associate it with certain, well-specified processes, the mathematical algorithms for which produce results that are at odds with what one would expect in the presence of any of the existing battery of genetic transmission processes. These processes are the population-level consequences of individual decision-making as we described above in the section "Intent Is an Appropriate Subject of Study and Component of Theory." There is simply every empirical reason to think that these processes are important in the case of cultural evolution. They have much in common with mate choice. They have much in common with virus susceptibility: We don't decide to catch a cold, but our immune system attempts to filter out all transmitted microbial pathogens, while attempting to remain susceptible to such beneficial microbes as symbiotic gut bacteria. Debating concepts like these at high levels of abstraction, if that is all we want to do, is an enterprise best left to postmodernists and their ilk.

Real scientists make models, collect data, and solve problems. Evolutionary archaeology will succeed to the extent that it taps the power of population thinking to do these things in its particular domain. We should remember that we share more ideas than we differ on. Science is basically a cooperative endeavor in which we divide that labor of the task and coordinate our individual activities, thereby achieving things that no individual alone could hope to accomplish. Like all human social activities, the collective project is accomplished by an immense amount of individual competition and rivalry. In the long run, however, successful institutions are the ones that somehow harness the individual rivalry to the larger social goal.

REFERENCES

Bettinger, R.L., 1991, *Hunter-Gatherers: Archaeological and Evolutionary Theory*, Plenum Press, New York.
Brandon, R.N., 1990, *Adaptation and Environment*, Princeton University Press, Princeton.
Cavalli-Sforza, L.L., and Feldman, M.W., 1981, *Cultural Transmission and Evolution: A Quantitative Approach*, Princeton University Press, Princeton.
Darwin, C., 1859, *On the Origin of Species by Means of Natural Selection, Or, The Preservation of Favored Races in the Struggle for Life*, Murray, London.
Geier, R.N., 1988, *Explaining Science: A Cognitive Approach*, University of Chicago Press, Chicago.
Kuhn, T.S., 1962, *The Structure of Scientific Revolutions*, University of Chicago Press, Chicago.
Mayr, E., 1982, *The Growth of Biological Thought*, Harvard University Press, Cambridge, MA.

Index

Abbott, A.L., 224
Abstruse mathematics, 68
Academic communities, 44
Accessibility, 197, 211, 213
Accidents, 151
Accomodation, 173
Acheson, S. R., 98, 99
Acheulian biface tradition, 190
   technology, 174
Action, 50, 83, 166, 172
   control of, 173
   course of, X, 109
   nexus of, 175
   power of, 175
Actor/s, 117
   ecological, 184
   -groups, 138
Adaptation, 10, 35, 70, 93, 140, 157, 159
   fine-grained, 213
   theories of, 93
   See also Culture, Groups, Information, Process/es
Adaptationist, 140
   behavior-regulatory, 183
   contemporary, 143
   perspectives, 26
   psychological, 184
   stories, 26
   scenarios, 27
   See also Approach, cultural-evolutionary
Adaptive
   change, 38
   differential, 39
   directionality, 34
   functional patterning, 144

Adaptive (*cont.*)
   innovation, 152
   interpretations, 143
   just-so stories, 225
   landscape, 147
   markers, 155
   radiation, 44
   response, 40
   role, 159
   specializations, 207
   strategy, 50
   success, 155
   system, 35
   topography, 149
   See also Forces, Processe/s, Trait/s, Variants, adaptive
Adaptiveness, 70
Adult, 154
Aesthetic, 53
Affect-laden
   emblematic symbol systems, 155
   ideological systems, 160
Africa, 151
   East, 205
African/s, 151
   Homo erectus, 192
Agency, 11, 44, 45, 167, 177
   attribution of, 166, 168, 169
   concept of, 168
   host, 49
   in evolution, 165
   limited intrinsic, 49
   primary, 46
   social, 187
   See also Human, Power

Agent/s, 166, 227
  conscious, 44
  of selection, 166
  organism as, 168
  social, 190
  See also Human
Aggregate-based theory, 91
Aggressive threat, 183
Agricultural wave, 153
Agriculture, 151
  maize, 225
Aiello, L.C., 190–193
Alexander, R.D., 91, 93, 155, 171, 229
Algae, 45
Algorithm/s, 75, 78, 79, 83
  behavioral, 64
  code, 73
  Darwinian, 8, 207, 210
  mathematical, 230
  quantitative, 229
Alleles, 145, 152
Alliances, 25, 209; *see also* Political alliances
Allomemes, 112–117, 119, 120, 125
Altruism, 93
  reciprocal, 93
Amazon, 101
American archaeology, 19, 25, 133
American Philosophical Society, 33
American scholars, 5
Ames, K.M., 97, 98, 110
Ammerman, A.J., 153
Amphibious vertebrates, 43
Analogies
  biological, 230
  use of, 229
  verbal, 230
  See also Artificial selection, analogy
Anatolia, 153
Anatomy, 190, 223
Animal/s, 45, 46, 157, 168, 204, 205, 210, 211
  carved, 198
  perforated teeth, 198
  tracks, 201, 213
Anthropocentrism, 177
Anthropoid genera, 192
Anthropology's mythology, 90
Anthropomorphic concept of selection, 166
  sense of volition, 166
Anti-Darwinian sentiment, 222

Antler handle, 124
  spoon, 124
  technology, 212
Apes, 201–203
Apiarist, 46
Apprenticeship, 54, 55, 174
Approach, cultural-evolutionary, 20
  Darwinian, x, xiii, 4–6, 8, 11, 61, 83, 110, 128, 172, 175, 193
  deductive-nomological, 20, 22
  ecological/functionalist, 5
  essentialist typological, x
  evolutionary, 13, 17, 222
  evolutionary ecological, 8
  experimental, 177
  group adaptionist, 7, 9, 10
  mathematical, 87
  neo-Darwinian, x, 43, 44
  qualitative, 87
  scientific evolutionary, 18
  selectionist, 10, 17
Archaeo-faunas, 142
Archaeological
  consequences, 159
  interpretation, 193
  language, 119
  phenomenon, 159
  See also Model
Archaeologist as culture historian, 141
  as paleopsychologist, 141
  evolutionary, 221
  -turned-philosopher, 20
Archaeology convention, 114
Archetypes, 51, 55
Architecture, brain, 171
  modular cognitive, 212
  psychological, 183
Ardener, S., 80
Argument
  good of the species, 47
  functional, 25
  inertia, 93
  optimality, 225
  teleological, 226
  See also Processual
Aridos, 205
Art/s, 5, 134, 159, 212
  expressive style, 155
  See also Rock art site, Style
Arthur, W.B., 149

INDEX

Artists, 199
Artifact attributes, 119
  ceramic, 53, 55
  eusociality, 51, 56; see also Eusocial animal
  form, 118, 165
  Indian behind the, 79
  -production system
  survival-promoting, 50
  types, 38
  utilitarian, 143
  worker, 53
Artificial selection, 168, 170, 223
  analogy, 166
Assemblage, artifact, 52
  cultural, 45
  of cultural features, 7
  of culture signals, 70
  ecological, 44, 46, 50
  faunal, 213
  lithic, 213
  of material signal, 64; see also Signal assemblage
  multiobject, 54; see also Eusocial anumal
  of pottery, 51, 53, 54
  of stone artifacts, 56
Assimilation, 173
Astronomy, 177
  Copernican, 223
  Ptolemaic, 223
atl atl, 124
Audience, 121
Australian Aboriginal settlements, 65
Australian prehistoric archaeology, 44
Authority, 95, 101, 116

Backed and self bows, 138
Bahn, P., 199
Baldwin, J.M., 168; see also Effect/s
Bandwagons, 19
Barkow, J.H., 102
Barth, F., 148
Barton, N.H., 157
Batammaliba houses, 62
Bateson ban, 75
Bateson, P.P.G., 83
Battleship curves, 148
  (lenticular) pattern of increase and decrease, 158
  of frequency seriation, 119
  of stylistic seriation, 147
Baumhoff, M.A., 159

Beads, 198
  ivory, 198
Beehive, 54, 55
Beer, G., 166, 167
Bees, 46, 52, 207
  worker, 52, 53, 55, 56; see also Domestic eusocial insects
Bednarik, R.G., 198
Behavioral combinations, 155
  ecology, 6
  See also Algorithm/s
Benefits culturally defined, 118
  biological, 118
  long-run, 150
Bettinger, R., X, 8, 110, 149, 159, 170, 172, 176
Biface to flake-based technology, 34, 35, 37, 40, 224
  producion, 36
  reduction, 36
Bigmen, 95
Bilateral symmetry, 124
Binford, L.R., 5, 9, 20, 92, 93, 137–140, 159, 190, 205, 206, 212
Biochemistry, 68
Biology is destiny, ix
Bird coloration, 143
Birth rates, 172
Bison, 206
Blind natural selection, 168
  watchmaker, 165, 179
Boas, F., vii
Boeda, E., 213
Boesch, C., 202
Bone, 125, 212
Bonner, J.T., 81
Boulders, 49
Bourdieu, P., 128
Boundary conditions, 22, 23, 70
  of cultural replication, 83
  See also Style
Bow and arrow, 99
Bowler, P.J., 169
Boyd, R., 7, 8, 91, 110–112, 114, 125, 127, 128, 143, 150, 151, 154, 155, 157, 170
Braidwood, R.J., 20
Brandon, R.N., 25, 226
Breeders, 52; see also Population/s
Bride-price, 157
Bridging arguments, 20
  law concept, 22
British Columbia, 125

Broca and Wernicke area homologues, 186
Brood, 52
Brothers, L., 187
Bruyere, B, 65
Buffering role of society, 175
Bugos, P.E., 94
Building/s, 64, 65, 74, 78, 79, 82, 171
Bureau of American Ethnology, 25
Burian, R., 25
Butterfly effects of dynamical systems theory, 81

Campbell, D.T., 91, 150
Cambridge Paleoeconomy School, 5
Cannon, A., 100
Cantabrigean Tradition, 44
Car, 157
Carbon atom, 23
Carlson, R.L., 126
Carr-Saunders, 90
Carrying capacity, 89, 91, 97
   environmental, 90
Cartesian method of reductionism, 168
Cartesian mind–body dualism, 170
Cartesian nonagency, 168
Cartesian program, 167
Carved
   human figures, 198
   pumice figurine, 123
   See also Animal/s, carved
Cartesian
   method of reductionism, 168
   mind–body dualism, 170
   nonagency, 168
   program, 167
Case, R., 212
Caste, 52, 53
Catastrophic cascades, 81
   mortality profiles, 206
   selection, 120
Catastrophies, 169
Catchment area, 55
Cattle, 139
Cavalli-Sforza, L.L., 148, 151–153, 230
Cave walls, 202
Celibate, 54, 82
Cell nuclei, 110
Central processes, 208
   system, 211
   tendency, 190, 193
Centrifugal force of competition, 175
Centripetal force of social integration, 175

Ceramic sociology, 223
Ceteris paribus, 148
Chagnon, N.A., 94, 95
Chaikin, I., 123
Chain of nodes, 179
Chaines operatoire, 174, 213
Chaotic dynamics, 144
Chance
   nonimitation, 146
   stylistic variation, 147
   See also Effect/s
Characters
   adaptive, 155, 156
   indicator, 157
   preference, 157
   symbolic, 156
   See also Display, Functional behavior, Male/s
Charisma, 47
Chase, P., 198, 199, 203, 206
Cheney, D.L., 202, 203, 207, 210, 211
Chiefdoms, 87, 94, 100, 102
Child, 178, 204
   development, 212
Childe, V.G., 20, 26
Children, 80, 154, 172, 207
Child's play, 227
Chimpanzees, 186, 203
Chinook, 122
Choice, 116, 127, 144, 160, 170, 172
   cultural, 119
   free, 115, 133, 134
   freedom of, 175
   individual, 150
   random stylistic, 147
   rational, 138
   selection-derived rational, 156
   See also Human, Model, Problem, inverse
Chomsky, N., 73, 186, 207, 208
Christ, 46
Circuitry
   dedicated, 186, 189
   lateralized, 186
   social cognitive, 187
Circumscription
   environmental, 92, 97
   social, 92, 95, 97
Civilization, 178, 179
Civilizing process, 178
Clans, 96
   high-status, 97
   See also Conical clan political organization

INDEX

Class, 113
  kinship, 202
  middle, 101
  socioeconomic, 113
Classification, 200–202, 206, 213
Climate, 37
Climatic gradients, 153
Cloak, F.T., 170, 171, 177
Clocks, 176–179
Coalition formation, 183
Coast Salish, 122, 125
Code algorithms, 73
  cultural, 73, 80, 82, 112
  elementary, 71; *see also* Recipes
  gene, 75
  generative, 73
  genetic, 73
  spatial, 77
  universal consistent, 83
Coercion, 156
Coevolution, 109, 113
Coevolutionary theory, 110, 117, 125–128
Cognition, 185, 187
Cognitive
  Darwinian archaeology, 8
  domains, 207, 213
  -developmental psychology, 185
  -emotional development, 189
  hierarchy, 212
  social skills, 188
  threshold, 212
  See also Modularity, Module/s, Process/es
Cohen, J., 75, 154, 179
Coiling, 54
Coke, 33
Cole, D., 123
Collecting, 175
Collective project, 231
  representation, 90
Combe Grenal, 206
Commodities, 100
Common ancestral production system, 128
  ground, 221, 222
  history, 119
  production systems, 119
Communication
  emblematic, 139, 153
  expressive, 139, 153
  modes of, 80
  parallel channels of, 67

Communication (*cont.*)
  skills, 190
  stress, 70
  systems, 73
  verbal, 65
  See also Intentional activity
Communicative elements of style, 153
Competitiveness, 91
Competition, 43, 92, 100, 101, 117, 154, 169, 170, 175, 221, 231; *see also* Power, Process/es, Status
Complexity, 100
Computational capacity, 209
  load of social relationships, 190
Conditional reasoning, 188
Conflict, 99–101, 177
  of interest, 150
Conformist transmission bias, 153
Confounding, 155
Confrontation
  direct, 169
Conical clan political organization, 150
Conjectures, 172
Conscious, 168
  activity, 166
  See also Agent/s
Conspecifics, 111, 115, 120, 201–203, 205, 211
Contingency, 12, 24; *See also* Laws
Context
  archaeological, 62
  Aurignacian, 198
  -dependent meaning, 193
  individual, 138
  Lower Paleolithic, 202
  material, 80
  Middle and Lower Paleolithic, 198
  natural, 202
  of engagement, 176
  selective, 205
  social, 11, 114, 121, 125, 138, 173, 175, 188, 201
Contextual analysis, 62
  view, 79
Contextualists, 80
Cooperation, 150, 228
  social, 188
Cooperative behavior, 227
Coordination, 228
Copernican system, 223
Copernicus, 177

Core-generalized, 36
  preparation, 174
  reduction, 36
Corporate dwelling, 100
  group, 95, 96, 99, 100
Corporatist political sentiments, 47
Corpuscles of culture, 171
Cosmides, L, 102, 135, 188, 189, 207, 210
Costall, A., 165, 169
Costs and benefits, x
Counterfunctional, 134
Coupland, G., 97–100
Craft specialists, 96
Craft specialization, 127
Croes, D.R., 97, 125
Cues
  auditory, 203, 210
  primary visual, 203
  secondary visual, 203, 210
  visual, 210
Cullen, B.R.S., 8, 12, 45, 48
Cultural, 109
  adaptation, 82
  change, 165, 171, 176, 213
  collapse, 81
  contact, 126
  degeneration, 78
  ecology, 7
  evolution, 91, 112, 230
  evolutionary processes, 144
  generation, 141
  hereditary material, 44
  hitchhiking, 153
  inheritance, 111
  instruction, 170
  mechanisms, 82
  message, 83
  phenotype, 49
  replication, 62, 68, 83
  reverse transcription, 50
  revival, 122
  selectionism, IX, XIII, 6–9, 43, 44, 45, 51
  selectionist theory, 110
  tradition, 141
  traits, 48
  viral phenomenon, 48, 50
  generated hypothesis, 52
  See also Determinism, Features, Inheritance, Innovations, Model, Pattern/s, Process/es, Replication, Transmission, Trial-and-error

Cultural reason, 157
Cultural science bias, 190
Cultural selection by imposition, 117
Cultural virus theory (CVT), 6, 43–47, 51, 55, 56
Culturally transmitted determinants of behavior, 150
Culture
  as adaptation, 140
  capacity for, 183
  carriers, 112
  change, 4–6, 11, 12, 83, 87, 102, 113, 115, 137
  -historical, 18
  -history, 20, 136
  process, 19, 20, 141
  products of, 176
  See also Distributed culture, Laws, Material
Curtiss, S., 188

Dancing, 78
Darwin, C., VII, 11, 25, 134, 142, 165–167, 169, 170, 222, 224, 226, 229; *see also* Tangled bank
Darwinian account
  archaeologies, 5, 7, 8, 11
  archaeology, ix, xi, 4, 6, 9–12, 127, 128; *see also* Cognitive
  concepts, VII, IX
  credentials, 5
  culture theory, 43
  enterprise, 222
  evolutionary archaeology, 17
  ideas, i, ix
  inference, 190
  mantle, 5
  methods, 222
  model of evolution, vii, 61, 69
  perspective, vii, 4
  population, 44
  psychology, 223
  schizophenia, 50
  thinking, 141
  umbrella, 5
  See also Algorithm/s, Approach, cultural-evolutionary, Explanation, Mechanisms, Psychology, Transmission
Darwinian theory, 68, 159, 165, 166, 169, 172, 228
Darwinism, 160, 193, 210
  neural, 44
  universal, vii

Davis, W., 119, 125, 127, 197, 198, 201, 202
Dawkins, R. vii, 9, 25, 43, 44, 46, 48, 52, 69, 73, 75, 82, 165, 168, 170, 177
Deacon, T.W., 186, 212
Death, 176
 rates, 172
Debate
 hunting/scavenging, 205
 neutralism controversy, 144, 146
 parameter values, 146
 sexual selection, 142
Decalage, 212
Decision
 political, 95
 power, 96
 rules, 150
 social, 95
 See also Power
Decision-making, x, 10, 68, 83, 91, 109, 115, 138, 160
 costs, 151
 processes, 87, 92, 97
 See also Individual/s
Decorated objects, 126
 spoons, 125
Decoration
 personal, 198
Decorative elements, 138
 motifs, 120, 125
Deetz, J.F., 143, 153
Deir el Medina, 62, 63 (illus.), 65, 71, 73–75
Deiocentrism, 177
Deity, 51, 167
Delluc, B., and G., 198
Dennett, D., 202
Dental reduction, 168
Dependence, 48, 50
Depiction, 197, 200
Deprivation, 102
Derrida, 45, 47
Descartes, 168, 178
Descent, 94
 with modification, 24, 119
 shared, 119
Design problems, 149
 research, 223
Desmond, A, 78, 79
Dethlefsen, E.S., 143, 153
Deterministic clockwork, 178

Determinism, 23
 biological, vii, 110
 blind, 166
 cultural, vii
 empirical, x
 environmental, 38
 genetic/memetic, 177
 social, vii
 technological, 177
 See also Laws
Dialectics of society, 176
 with the past, 176
 social, 175, 176
Dibble, H., 199
Diffusion, 112, 126
 rates, 147
Disease, 123
 of the mind, 46
Directionality, 179
Directional selection, 120
Disequilibrium, 65
Display, 117
 characters, 157
 public, 111
 stylistic, 139
 symbolic, 157
Disposable goods, 55
Dissonance, 65, 81
Distributed culture, 192
Distribution
 of cranial deformation, 122
 critical resource, 92
 labret wear, 122
 polymodal, 74
 probability, 71
 spatial distance, 75
 See also Egalitarian
Distrust, 227
Divergence, 147
Diversity, 82
Divine intervention, 3, 166, 167
 watchmaker, 178
DNA, 48, 50, 51, 78, 79, 110
Dog, 224
Domains of activity, 213
Domain-specific abilities, 185
 cognitive processes, 211
Domestic eusocial insects, 56
 life, 51
Domesticates, 45, 46, 50, 56, 166

Domestication, 26
Domesticator, 50
Domination, 12
Drift/s
　chance, 146
　cultural, 112, 119, 126
　to fixation, 146
　See also Genetic/s, Selection, Statistical association
Drosophila, 145
Drones, 52, 54
Drying, 54
Dual inheritance theory of coevolution, 109–111, 113, 115, 116, 118, 119, 122, 127, 128
Dual inheritance theory of style, 121
Duff, I.A., 198–200
Dunbar, R.I.M., 190–193
Dunnell, R.C., 6, 26, 113, 118, 127, 135–139, 141
Durham, W.H., 7, 8, 43, 109–116, 127, 128, 172
Durham's theory of coevolution, 127
Durkheim, E., vii, 91
Dvorak keyboard, 33, 144

Early Late Phase, 99
Eastern Woodlands, 35
Eberhardt, W.G., 143
Ecological anthropology, 92, 93
Economic
　control, 95, 100
　monopolies, 94
　production, 95
　properties, 96
　redistribution, 95
　success, 115, 157
　system, 157, 193
　See also Revolution
Economics of nature, 115
Economists, 149
Economizing measures, 224
Ecosystems evolution, 91
Edelman, G.M., 75, 83
Effect/s
　active hitchhiking, 156
　adaptation-generating, 143
　Baldwin, 168
　chance, 147

Effect/s (cont.)
　decision-making, 143
　drift-enhancement, 151
　fitness, 144
　frequency-dependent, 143
　hitchhiking, 153, 157
　random, 145
　runaway, 157
　selection-derived, 143
　See also Variation, bias, Butterfly efforts of dynamical systems theory, guided
Effective
　number of influential investigators, 148
　sizes of populations, 149
Egalitarian, 95, 99
　distribution of grave goods, 150
　See also Gender egalitarianism
Egypt, 62
Einstein's theory of relativity, 223
Eldredge, N., 44, 115
Elective affinities, 167
Elias, N., 178
Elites, 96, 118
Elton's Gridlock, 80
Embryonic development, 24
Empiricism, 21
　prima facie, 228
Endocranial volume, 191
Engineering–design analysis, 26
Environmental
　carrying capacity, 90
　degradation, 92
　determinism, 38
　improvement, 92
　See also Circumscription, environmental
Engineers, 149
Environment
　spatially heterogeneous, 151
　See also Evolutionary, Pleistocene
Epigenetic naturalism, 172
E.Q., 192
Equifinality, 158
Equilibrium, 65, 149; see also Multiple evolutionary equilibria
Esoteric lore, 148
Essentialism, 18
Ethical norms, 228
Ethics, 188
Ethiopia, 52

INDEX

Ethnic
  boundary, 160
  groups, 113, 121, 154, 159
  identity, 154
  isolate, 160
  marker, 121
Ethnicity, 120, 121, 127
Ethnoarchaeology, 22, 159
Ethnographic analogy, 22
  data, 199
  description, 36, 205
  record, 101
Ethnolinguistic unit, 228
Eusocial animal, 56
  artifact assemblage, 56; see also Artifact attributes, eusociality
  colony structures, 52
  hereditary structure, 56
  object assemblages, 56
  pottery, 56
Eusociality, 51
Euroamerican, 122
Eurocanadian, 122
Europe, 153
  Middle Pleistocene of, 205
  See also Upper Paleolithic, record of Europe
European contact, 125
  diseases, 123
  dominance, 123
Evolution
  artistic, 156
  biological, 3, 8, 10, 11, 112, 113, 115
  of the capacity for visual symbolism, 200
  cognitive, 8, 193, 197, 211
  cultural, 8, 11, 26, 109, 110, 114, 120, 128, 149, 150, 221, 227
  dialect, 156
  directional, 167
  ecosystem, 91
  genetic, 8, 159
  human cognitive, 185
  of ideological and ceremonial systems, 154
  of language, 153, 154
  of the mind, x
  neutral and selective, 146
  random, 146
  of science, 148
  scientific, 19

Evolution (cont.)
  social, 129
  of social behavior, 93
  of social systems, 190
  of stylistic cultural variation, 145
  of stylistic features, 144
  See also Agency, Explanation, Hominids, Modes of power, Process/es, Tower of Babel evolution
Evolutionary
  archaeology, 8, 24, 26, 27, 119, 223
  biologists, 7, 90, 93, 134, 142, 157
  biology, 44, 47, 48, 51, 52, 56, 87, 102, 141, 143, 147, 183, 222, 223
  culture theory, 223
  ecologists, 10, 87, 136
  ecology, xiii, 7, 224; see also Optimal behavior
  environment, 210
  fact, 3, 4
  history, 34, 102
  models of stylistic variation, 135
  processes, 34, 150, 158, 221
  psychology, XIII, 3, 12, 183, 193, 227
  research, 225
  sequence, 179
  synthesis, 27
  theory, 3–6, 12, 18, 23, 160, 208, 210, 223, 227, 228
  See also Forces, Mechanisms, Process/es
Exchange, 92, 93, 98
  partners, 150
  primitive, 94
  social, 186, 189; see also Rule/s
Exemplars, 223
Experimentalists, 27
Explanadum, 22
Explanation, 28
  archaeological, 158
  causal and materialist explanations of social phenomena, 158
  Darwinian, 10, 11, 43
  evolutionary, 35, 40
  Hempelian, 23
  how actually, 226
  how possibly, 226
  teleological, 167
  See also Scientific explanation
Exploitation, 12

Face
  emotional expressivity, 187
  processing, 187
  -specific area, 187
Faith, 46, 47, 223
Features
  arbitrary stylistic, 155
  cultural, 64, 83
  stylistic, 147, 152, 155
Feldman, M.W., 148, 151, 152, 230
Female/s, 188, 191
  adult, 192
  choice, 142
  preference, 142
Fergusson, E.S., 179
Fetishism of commodities, 177
Fibonnacci series, 73
Fire wall, 227, 228
Firing, 54
Fitness, 6, 35, 38, 172, 224
  artifact, 15
  cultural, 115
  Darwinian, 118
  genetic, 115
  function, 228
  inclusive, 93, 116, 184
  of a potential mate, 134; *see also* Handicap hypothesis
  payoffs, 190
  reproductive, 109
  See also Values, cultural
Fittest, 166; *see also* Survival of the fittest
Fixation, 119
  by drift, 149
  of belief, 208
Flannery, K.V., 19, 25
Fletcher, R.J., x, 12, 71, 80
Flinn, M.V., 155
Flint knapping, 174
  Florentine court of the Medicis, 47
Fodor, J., 186, 207, 208, 211
Food preferences, 114
  processing, 168
  processing factories, 125
Footprints, 205
Forager, 204, 205
Foraging, 183, 189
  activity, 213
  behavior, 210
  efficiency, 206
  See also Pleistocene

Forces
  adaptive, 153
  conveyance, 112, 114, 126
  cultural evolutionary, 112
  directional evolutionary, 135, 153
  evolutionary, 120, 122, 128, 141, 144
  nonconveyance, 112, 126, 127
  random evolutionary, 135
  related evolutionary, 135
  vicarious, 150
  See also Selective advantage, Transmission
Form
  house, 151, 152
  -lines, 122, 124
  residential, 125
  sculptural, 122
  split U-, 124
  technical, 148
  vessel, 120
  See also Artifact attributes
Foucault, M., 168
Founder effect, 44
Founding fathers, 90
France, 33
  southwest, 198, 206
  western, 198
Fraser Plateau, 101
Freemasonry, 154
Free will, 133, 134, 166, 170
Freezing of a gesture, 201
Frequency-dependent
  biased cultural transmission, 151, 152
  selection, 228
  See also Rule/s
Friends, 150
Fritz, J.M., 23
From dams to refrigerators, 149
Function/s, 119
  communicative, 153
  mechanical, 74
  memory, 187
  pot, 120
  social, 200
  See also Power
Functional behavior, 133
  characters, 139
  efficacy, 53
  interpretation of behavior, 11, 12
  response, 224
  variation, 145
Furnishings, 150

Galileo, 47, 177, 178
Galton's problem, 153
Gamble, C., 134, 205
Games, 150; see also Prestige
Gardner, K.E., 80, 207–211
Geier, R.N., 223
Geissenklösterle, 198
Gender egalitarianism, 160
Geneological independence, 48
General
 -purpose information processor, 207
 -purpose learning mechanism, 207
 systems theory, 19, 25
 See also Intelligence
Generic individuals, 9
Genetic/s, 143
 drift, 145, 147
 loci, 145
 material, 73
 molecular, 145
 processes, 56, 61
 program, 172
 See also Showy-tail gnus
Genes
 adaptive, 152
 frequency, 223
 neutral, 152
Geneticists, 27; see also Experimentalists, Population/s
Genie, 187, 189
Genome, 170
Genotype, 49, 50, 113, 115
Geocentric/deiocentric Christian model of the universe, 177
Geographic isolation, 27
Germany, 198
Ghana, 71
Ghost of Christmas Past, 167
Giant yams, 157
Gibbon song, 187
Gibson, K.R., 187, 212
Gillespie, J.H., 145
Glaciers, 49
Glenrose Cannery site, 124
God, 82, 178
Gold, 125
Golinski, J., 47
Goodnow, J.J., 175
Gorillas, 191, 203
Gormandizing, 97

Gould, S.J., 35, 44, 70, 75, 120, 212, 225
Gradualism, 27
Grammar Acquisition Module, 188
 verbal, 64, 73
Grammatical structures, 207
Graves, 150; see also New England gravestone styles
Graves-Brown, P., 11, 12, 172
Gravity, 167, 177
 fixed law of, 225
Great apes, 187
Great Basin, 160
Great Plains, 35
Greene, W.C., 50
Greenfield, P.M., 212
Griesmer, J.R., 148
Grene, M., 68
Group/s
 adaptation, 91
 band size, 190
 behavior, 6, 92
 corporate economic, 100
 emblematic information, 154
 lineage-based corporate, 102
 membership, 154
 mind, 90
 preservation mechanisms, 90
 population, 92
 reference, 112–117, 128
 selection, 90, 173, 228, 229
 subordinate reference, 116
 subreference, 116
 See also Ethnic, Nuclear family group
Gut bacteria, 230

Habitas, 114, 128
Habitats, 156
Hackenberger, S., 97, 125
Hahn, J., 198
Haida, 98, 99
Hamilton, W.D., 93, 116, 143, 229
Handicap hypothesis, 134
Harcourt, A.H., 191
Hardware, 150
Harris, M., 20, 25
Hauser, M.D., 187
Hayden, B, 97, 100
Headmen, 97, 99–101
Hefner, H.E. and R.S., 186
Hegel, 178

Hegelian theory, 178
Hempel, C., 20, 21
Hempelian notion of science, 21, 22
Hereditary inequality (material, origins), 95
  nobility, 96
  social inequality, 7, 90, 92, 94, 97, 100, 102
  status differences, 95, 101
Heritage constraint, ix, 12, 61, 64, 70, 77, 81
Heterozygous, 145
Hierarchical
  integration of executive structures, 212
  logic, 68
  mental construction skills, 212
  model of adaptation, 81
Hierarchies copying, 78
  political, 92
  replicative, 78
  social, 92, 98, 169
Hierarchization of cognitive evolution, 197
  cognitive processes, 201, 212, 213
Hinde, R., 8
Historians, 159
Historic contact, 96
  period, 4
  process, 24
Historical trajectories of change, 137
Hitchhiking, 152
  active, 155, 160
  passive, 155
  See also Cultural, hitchhiking, Effect/s, Hypothesis
Hobbesian "war each against all," 169
Hodder, I., 62
Hohlenstein-Stadel, 198
Hoko River, 124
Holland, T., 8, 10, 23, 24, 27, 70, 119, 120
Holm, B., 122, 124
Home range area, 192
Hominid/s, 201, 202
  encephalization, 190
  evolution, 192
  extinct, 190
  fossil, 191
  group size, 193
  Late Pleistocene, 168
  Lower and Middle Paleolithic, 198, 213
  social systems, 190, 193
  See also Pleistocene

Homo, 193
  erectus, 190, 192
  habilis, 192
  sapiens sapiens, 67, 189, 190, 192
Honey pots, 52
Hoofprints, 204
Hooker, J., 47
Horses, 153
Horticulturalists, 158
Horticulture, 37, 95
Hoxne, 205
Human
  agency, 5, 10, 54, 55
  agents, 45, 56, 170
  as intent-driven, 25
  behavior, 34, 39, 90, 102, 111
  choice, 10, 170
  communities, 53, 56, 68, 69, 79, 83
  consciousness, 44
  epigenesis, 173
  genotypes, 109, 111
  individual action, 44
  initiative, 54
  invention, 26, 170, 176
  knowledge, 178
  perceptual accuracy, 74
  phenotype, 9, 24, 34, 111; *see also* Inclusive phenotype position
  stylistic variants, 146
  symbolic systems, 154
  See also Foraging, Power, Symbols
Humphrey, N.K., 170, 202, 209
Hunter-gatherer/s, 204, 228
  ethnographically documented, 199
  mobile, 80
  prehistoric, 9, 5
  See also Rock art site, Village-based political units
Hunters, 160
Hunting, 95, 175
  behavior, 201
  herd, 206
  of large game, 205
  selective, 206
  specialized bison, 206
Huxley, J.S., 27
Huygens in The Netherlands, 177
Hyper-Machiavellian, 188
  males, 189

INDEX

Hypothesis
  adaptive, 158
  hitchhiking, 153
  neutralist, 145, 146
  Parry–Kelly, 224
  processual, 152
  Renfrew's, 153
  runaway, 134, 142, 157, 158; *see also* Counterfunctional
  selectionist, 145
  signaling, 157
  Wynne-Edwards, 90
  See also Wright, S.

IBM 370, 149
Iconic, 204
Identity, 138
Ideological phenomena, 110
Ideology, 159, 160
  Christian, 46
  of Christianity, 177
  supernatural, 134
  See also Western ideology of objects
Idiot savants, 209
Igloos, 50
Ilskaya, 206
Image making, 197, 200
Imposed cultural selection, 116, 128
Imposition, 127
Inanimate secondary visual cues (ISVCs), 203–206
Inbreeding avoidance, 183
Inclusive phenotype position, 43, 44, 48
Indeterminancy, 193
Indexes, 203
Individual/s
  acted-as-if, 225, 226
  adaptation, 7
  decision-makers, 184
  decision-making, 175, 227, 230
  intent, 226
  intention, 228
  invention, 152
  motivations, 87
  prestigious, 156
  self-interested, 227
  self-sacrifice, 228
  status striving, 94
  See also Human, Innovations
Individualistic perspective, 8, 9

Indo-European, 153
Industrialized societies, 80
Infant, 172
  dependence, 173
Information
  inheritance, 110
  packages of cultural, 113
  -processing adaptations, 183
  processing device, 209
  socially transmitted, 230
  systems, 112
  unit of, 113
  See also Process/es, Transformation
Influence, 148; *see also* Power
Influential investigators, 148
Inheritable social inequality, 94, 100, 102
Inheritance
  biological, 69, 87
  cultural, 120, 227
  cultural mode of, 87
  culture as a system of, 140
  genetic, 120
  modes of, 7
  of variation, 143
  See also Mendel
Inherited characteristics, 168
  decision rules, 171
Initiation ceremonies, 199
  rites, 148
Innovations, 112, 126, 153, 156, 177
  cultural, 153
  individual-level, 153
  random, 158
  rates, 147–149, 152
  simultaneous, 146
  stylistic, 147
  technical, 148, 152, 160, 177
  See also Adaptive, Effective
Innovative, 176
Insect intromittent organs, 143
Institutions
  religious, 159
  symbolic, 159
  See also Slavery, institutionalized
Instrumentation, 223
Intelligence
  bodily-kinesthetic, 209–211
  generalized, 185, 211–213
  linguistic, 208, 211
  logical mathematical, 208, 210, 211

Intelligence (*cont.*)
　musical, 208
　nonsocial, 203, 207, 210, 211
　personal, 209, 211
　social, 203, 207, 209–211
　spatial, 209–211
　See also Multiple evolutionary equilibria, intelligences
Intensification of resource procurement, 92
Intent, 227, 230; *see also* Problem, inverse
Intentional activity, 168
　communication, 201, 202, 206, 213
Intentionalist account of natural processes, 169
Intentionality, 5, 10, 26, 167, 169
　second-order, 202
Interaction pressures, 70
　social, 185, 201, 203
Internal clockwork, 168
Interpersonal distances, 67
IQ-type test, 188
Isaac, G.L., 189, 193
Islander dialect, 154
Isochrestic, 139, 158
　behavior, 138
　choices, 138
　concept, 138
　cultural variants, 145
　variants, 138, 143, 146
Isochrestism, 120, 127
Isolation, 151
　social, 188
　See also Geographic isolation
Isolines, 153
Isotherms, 153

James, T.G.H., 65
Jarman, M.R., 5
Jefferson, T., 33
Jeske, R.J., 40
Jochim, M., 91, 92, 97
Jones, G.T., 224
Jurgens, U., 186
Just-so stories, 25, 226
　adaptive, 225

Kayaks, 50
Keeley, L., 98
Keesing, R., 111

Kelly, R.L., 36–39
Kenya, 52
Kilns, 54
Kimura, M., 145
Kin network, 95
　perspective, 94
　recognition, 210
　selection, 91, 93, 94
Kings, 52
Kinesics, 64
Kinship
　distance, 94
　relations, 211
　See also Class
Kinsmen, 91, 93
Kirkpatrick, S., 149
Klasies River Mouth, 206
Klein, R.G., 206
Konokomba village, 71
Knight, C., 47
Kroeber, A.L., 136, 136
Kuhn, T.S., 177, 178, 223
!Kung arrow points, 150
　camps, 65
　hunter, 204

Labor, 96
　division of, 173, 228
　sexual division of provisioning, 189
　social, 117
　women's, 160
Labov, W., 154, 156
Labrets, 125, 127
Lachane site, 124
La Ferrasie, 198
Lamarckian, 49, 75, 167, 227
　needs- or pressures-driven theory, 75
　response, 81
Lande, R., 157
Language/s, 8, 10, 65, 113, 153, 156, 160, 185, 197, 202, 207, 212
　guided reinvention of, 178
　of purpose, 168
　sign, 203
　skills, 186
　See also Structure, deep message
Large-scale social process, 117, 128
La Souquette, 198
Late Phase, 98
Later Stone Age, 206

Laws, 21, 22, 28
 chemical–physical, 23
 contingency, 23
 covering, 20
 of culture, 128
 of culture process, 12
 deterministic, 23
 empirical, 20
 of history, 129
 invariant, 23
Layton, R., 200
Learning, 10, 175, 227
 mechanism, 207
 social, 7, 87, 113, 229
Least-squares regression, 192
LeBlanc, S., 21
Leonard, R.D., 224
Leontiev, A.N., 174
Lethal systemic drugs, 50
Levallois technology, 174, 213
Levins, R., 158
Lewontin, R.D., 25, 171, 172, 173, 225
Liberal legislation, ix
Lichen, 45
Lieberman, P., 192
Life, 176
Lineage, 170
 areas, 101
 central, 53
 headman, 95
 highest-ranking, 100
 members, 97
 membership, 94
 powerful, 101
 segmentary, 150
Linguistic/s, 49, 64, 186, 228
 change, 156
 historical, vii
Linkage, 142
Literature, 78
Locals, 151
Lock, A., 178
Locke, J.L., 188
Locus of cultural selection, 126
Logical positivism, 21
London, 80
 Zoo, 52
Long-run, 231
 data, 159
 of rational result, 179
 See also Benefits culturally defined

Los Angeles, 80
Love, 46
Lower Columbia River, 122
Luck, 120; *see also* Variants, adaptive
Lukes, S, 116
Lumsden, E.J., 152
Lysenko, 169

Macaques, 186, 187
Machiavellian, 188; *see also* Hyper-Machiavellian
Machiavellianism, 188, 189
Mach V, 189 (illus.)
Mackinnon, J., 191
Magnitudes of selective differences, 149
Magdalenian, 174
Main settlement, 63 (illus.), 65, 66 (illus.), 71 (illus.), 73
Maladaptive, 101, 142
 behavior, 96
 decisions, 93
Male/s, 114, 148, 188
 -biased display characters, 157
 characters, 142
 deer, 204
 noble, 97
 reproductive success, 205
 undergraduates, 188, 189
Malthus, 169
Mammals, terrestrial, 202
Man, x
Manganese oxide, 198, 202
Mangelsdorf, P.C., 148
Manual dexterity, 8
Mark making, 200–202, 206, 211, 213
Markovian processes, 119
Marshack, A., 198
Martha's Vineyard, 154
Martindale, C., 155, 156
Marx, K., 177, 178
Marxism, 47
Maschner, H.D.G., 8, 98, 99, 101
Masks, 122
Material
 culture, 9, 46, 112, 120, 127, 173
 hereditary, 50, 53, 54
 spaces, 67, 78
 wealth, 95
 See also Raw material
Materialism, 18
 contemporary, 140
 historical, 129

Materialist
   account of evolution, 168
   cultural, 136
   strategy, 18
Mate, 100, 101, 134
   choice, 135, 183, 227, 230
   -choice sexual selection, 157
Mathematics, 209
   defense, 183
   See also Approach, cultural-evolutionary
Mating, 224
Matrilineages, 100
Matrilineal, 20
Matson, R.G., 98, 124
Maximizing creatures, 25
   strategy, 25
Mauran, La Cotte, 206
May, A.D., 80
Mayr, E., 24, 27, 28, 120, 228
McCay, B., 91
McHenry, H.M., 192
McNeary, S., 98
Mead, M., 173
Meaning attribution, 200, 206, 213
   capacity to attribute, 201
   inside, 199
   outside, 199
Means of production, 177
Mechanisms
   Darwinian, xiii
   evolutionary, 142
   of adaptation, 91
   sexual selection, 134
   supraindividual, 90
   See also General, Learning, Psychological consensus
Mellars, P., 198, 206, 212
Meme/s, ix, x, 48, 82, 110, 112–117, 119, 120, 125, 171, 176
   fertile, 170
Mendel, 68
Mendel's theory of inheritance, 166
Mentors, 157
Mesoamerica, 35
Metacognition, 187
   physical agency, 167
Methane, 23
Mexico, 148
Microbial pathogens, 230
   climate, 152
   dialectic change, 154

Microbial pathogens (cont.)
   evolutionary studies, 154, 155
   scale processes, 158
   time scale, 159
Middle Paleolithic: see Paleolithic
Middle Ages, 178
Middle East, 153
Middle, Late Phase, 99
Middle Phase, 99
Middle range studies, 159
Middle Stone Age, 206
Migrants, 151
Migration, 112, 126, 147, 152, 156, 159
Military success, 117
Miller, J., 45, 47
Millet, M, 117, 118
Mithen, S., 8, 12, 190, 193, 212
Mobility, 37, 38
Model
   archaeological, 189
   causal, 158
   central place foraging, 189, 193
   choice-based, 173
   coherence of a message recipe, 77
   cultural transmission, 173
   duel inheritance, x, xiii, 7, 8, 87; see also Hierarchical
   encephalization, 185
   epidemiological, 230
   evolution of visual symbolism, 197
   geocentric, 178
   group size, 193
   least-squares regression, 191 (illus.), 192
   logical and mathematical axiomatic systems, 21
   Man the Shoveler, 205
   mechanistic, 178
   multichannel, 68; see also Systems, biological
   of adaptive processes, 149
   of the evolution of sex, 142
   of speciation, 142
   Piaget/Popper, 171
   theoretical, 143
   sorting, 75
   virus, 69
   See also Evolutionary, Geocentric/deiocentric Christian model of the universe
Modern West, 155
Modes of power, 117
   of selection, 120

Modularity, 212, 213
 cognitive, 185
 mental, 197, 206, 207, 210
 position, 211
 strong, 185
 weak, 186, 189
Module/s, 208
 cognitive, 207
 mental, 198, 201, 207, 211, 213
 perceptual, 208, 211
 strong, 187
 visual perception, 208
 weak, 186, 193
Moiety, 97
Monitoring position, 128
Monkey/s, 187, 201
 vervet, 202, 203, 210
Monomorphic, 145
Morgan, C., 176, 177
Morgan, H.L., 25
Mortuary analysis, 223
Motherhood claims, 222
Motivation, 226
Mouldboard of Least Resistance, 33
Mountain Ok, 148
Mutation, 170
Multiple evolutionary equilibria, 149, 150
 intelligences, 207, 209
Munyimba, 71
Murdock, G.P., 90
Museums, 122, 123
Music, 136
Mutation
 random, 176
 rates, 145, 147
Mutationist view of cultural change, 177
Myopic optimizer, 149
Mystics, 226

Nagel, E., 20, 22
Naked mole rat, 52, 53, 55
 colony, 54
Native American societies, 122
Natural–cultural differences, 230
 similarities, 230
Naturalists, 27
Natural packages, 140
 signs, 197, 201, 204, 205
Natural selection
 intrasocietal, 114

Natural selection (*cont.*)
 intersocietal, 114
 See also Selection, Theology
Natural theology, 166
"Nature red in tooth and claw," 169
Near-decomposable, 176
Neiman, F., 147
Neocortex, 186, 190
 ratio, 191, 192, (illus.)
Neo-Darwinian, 229
 concepts, 133
 inheritance structures, 56
 logistics, 49
 umbrella, 222
 See also Approach, cultural-evolutionary
Neofunctionalists, 136
Neopositivists, 22
Netting, R.M., 101
Network externalities, 149
Neural circuits, 212
New archaeologists, 20, 23, 75, 136, 140
New Archaeology, 159, 223
New England gravestone styles, 148, 153
Newgrange, 176
New Guinea, 148
New Kingdome, 62; *see also* Settlements, of
 artisans
New Synthesis, 68
New York City, 123
Nepotistic tendencies, 94
Neuronal groups, 55
 group selection, 55
 path formation, 75
Neutrality, 35
Niska village, 98
Noble, 97
Nonhuman living systems, 167
 -primates, 186, 191, 202, 206
 -industrial societies, 95
 -processual, 68
 -reproductive individuals, 56
  worker objects, 53
 -social contract, 188
 -social knowledge, 210
 -symbol-using society, 201
 See also Intelligence, Symbols
Normal social environment, 186, 187, 189–193
Normative
 school, 140
 theorist, 140
 view, 141

North America/n, ix, 36, 37
  eastern, 40
  southwest, 35, 101
  transition, 224
Northern British Columbia, 98, 99, 124
Northern Northwest coast, 89, 96–101, 123
  art, 110, 118, 121–128
  inequalities, 98
  material, 123
  prestige system, 125
  system of social stratification, 125
  woodworking techniques, 125
Novice, 174
Nuclear family group, 189
Nucleated villages, 37
Nucleic acids, 229
Numic speakers, 160

Object/s, 166
  art, 134
  complex, 137
  utilitarian, 134
O'Brien, M., 8, 10, 23, 24, 27, 70, 119, 120
Occipital lobe, 187
Offspring, 191
Ohta,T., 145
Oligarchy, 117
  Roman, 118, 128
Omnivores, 114
Ontogenetic flexibility, 173
Ontogeny, 173, 187, 212
Ontology, 57, 144
Opler, M.E., 140
Optical illusions, 208
Optimal behavior, 93
  benefit, 50
  foraging theory, xiii, 6, 26
  just-so stories of evolutionary ecology, 225
  subsistence behavior, 155
Optimization texts, 149
Optimum, local, 149
Orangutans, 191
Oregon, 122–124
Origins of agriculture, 26
  art, 4, 206, 210
  big game hunting, 205
  hereditary social inequality, 87, 102
  hereditary status, XIII, 94
  species, 167

Origins of agriculture (*cont.*)
  ultra-sociality, 229
  visual symbols, 201, 202
Ovoids, 122

Paleoanthropologists, 159
Paleolithic, 4, 200
  early, 205
  Lower, 190, 213
  Middle, 206, 213
  Middle/Upper transition, 197, 212
  Middle and Lower populations, 199
  record, 192
  societies, 193
  See also Context, Hominid/s, Upper Paleolithic
Paley, W., 167, 178
Paradigm, 185
  completed, 223
  shift, 177
  See also Piagetian paradigm
Paradigmatic case studies, 223
Parasite, active, 170
Parasitic ecological relationships, 45
  phenomena, 46
  unintentional situations, 45
Parental generation, 146
Parents, 190
Parker, S.T., 187
Parry, W.J., 36–39
Pastoralists, 139
Patrilineage, 95, 101
Patrilineal, 20
Patron–client relationships, 117
Pattern/s
  cultural, 75
  drift-induced, 148
  empirical, 152
  long-term, 160
  macroscale, 158
  of language, 153
  of Marakwet residence, 62
  random, 144
  residential, 73
  spatial, 147
  Turing, 75
  "wired up" complex behavior, 187
  See also Settlement/s, of artisans
Patton, P., 33
Paul Mason site, 98, 99

INDEX

Peacocks, 157
Peckham, M., 167
Pelegrin, J., 174
Pendants, 198
Pender Island Site, 124, 125
Pendulum, 175, 177, 178
Perception, 197
   processes of, 208
Perceptual categories, 83
Period forcing, 81
Periodicity analysis, 74
Pfaffenberger, P., 177
Phenotype, 46, 50, 110, 112, 115, 120
Phillips, P. 19
Philosophical discourse, 118
Philosophy, Christian, 46
   of biology, 28
   of science, 20, 21
Phonology, 209
Phylogenetic heritage, 49
Phylogeny, 212
Physics, 177
Physiology, 223, 224
Piaget, 185, 209
Piagetian exploratory core reduction process, 174
Piagetian paradigm, 174, 175
Piagetian scheme, 173
Piaget's "Genetic Epistemology," 171, 172
Pianka, E.R., 91
Planets, 167
Plant/s, 45, 46, 168
   domesticated, 166
   storage, 160
   structure, 56
   style, 134
   See also Strategy, common reciprocating
Plateau, 101
Plato, 47
Platonic essences, 120
Pleistocene, 113, 145, 183, 184, 189
   early, 198, 205
   environment, 8, 186
   hominids, 202
   human foragers, 188
   Middle, 192
   psychology, 8
   Upper, 205
Plog, S., 23, 121
Polemic essays, 222

Polymorphic loci, 145
Political alliances, 97
   competition, 154
   complexity, 92, 94, 102
   correctness, ix
   economy, 117
   groups, 102
   unit, 94
Polymorphism, 145
Ponape, 157
Pool of information, 111
Popper, K.R., 171
Population/s
   ancestral, 156
   ancient, 190
   animal, 145
   biology, 228
   breeding, 228
   descendant, 156
   differentiation, 147
   ecology, 143
   effective size, 147, 148
   genetics, 143, 146
   geneticists, 144, 146
   in adaptation, 93
   -level behavior, 92
   -level consequences, 225, 226, 230
   thinking, 18, 142, 228
   pressure, 26, 89, 91
   raw size, 147
Postexcavational analysis, 51
Postmodernism, 193
Postmodernists, 133, 159, 230
Postprocessualists, 12, 62, 68, 79, 140
Poststructuralists, ix, 158
Potlach, 95, 96, 125
Potters, 54
Pottery industry, 54
   making knowledge, 55
   phenotype, 55
   style, 120
   wheels, 54
Povinelli, D.J., 187
Power, 12, 116, 117, 128, 159, 177, 193, 225
   active, 167
   autocratic, 95, 100
   competition, 183
   concept of, 115, 117
   exercise of, 115
   function, 191

Power (cont.)
    of decision, 170, 171
    of human agency, 168
    of population thinking, 230
    organizational, 117
    perceptive, 168
    positions of, 99
    structural, 117, 118
    tactical, 117, 118
    to control, 102
    to influence, 102
    See also Action, Decision
Practitioners, 148
Pragmatics, 209
Predation risk, 190
Predator alarm calls, 202
Prediction, 28
Predispositions, 193
Prehistoric assemblages, 36
    foraging, 6
    human actors, 102
    technologies, 34
    stone technology, 84
Pre-Numic, 160
Prescriptive, 111
Prestige, 12, 95, 155
    contests, 157
    games, 157
    system, 156, 157
Priests, 46, 47
Primate/s, 202
    anthropoid, 187, 191
    physiological ecology, 190
    taxa, 190
    See also Nonhuman living systems
Primatologists, 190
Prince Rupert Harbor, 98, 124, 125
Principal of altruistic sterility, 53
    inertia, 92
    parasitism, 45
Prisoner's dilemma, 227
Probabilistic regional survey, 223
Problem, inverse, 158
    of choice, 172
    of intent, 226
    solving, 10, 209, 222
Process/es
    adaptation-producing, 148
    adaptive, 148
    cognitive, 197, 198, 200, 205, 206, 213

Process/es (cont.)
    of competition, 169
    cultural, 140, 142
    of cultural change, 179
    of cultural evolution, 159
    of evolution, 143
    evolutionary perspective of culture, 143
    formation, 142
    interacting, 142
    large-scale, 225
    nonadaptive cultural, 148
    nonrandom, 140
    ontogenetic, 172
    of organic evolution, 221
    physical, 197, 200, 206
    random, 140
    of reinvention, 178
    selection, 140
    sexual selection, 143
    small-scale functional, 225
    See also Evolutionary, Hierarchical, Perception, Psychological consensus, Struggle
Process-related information, 159
Processual, 6, 35, 68
    account of cultural evolution, 159
    archaeology, 5, 7, 10, 12, 17, 19, 34, 40, 68, 82, 112, 129
    argument, 36, 37
Processualists, 139
Production of stone tools, 213
Projectile points, 35, 139
Prosopagnosia, 187
Protestant world, 177
Psychological consensus, 79
    familiar states, 82
    mechanisms, 207, 210
    processes, 211, 155
    propensities, 8
    research, 172
    species-typical, 183
Psychology, 91, 102, 174
    Darwinian, 223
    folk, 227
    See also Cognitive
Public goods scenarios, 227
Python, 203
    track, 210

Qualitative description, 5
Quality of life, 116

INDEX

Quantitative studies, 5
Quasi-isolation, 44
Queen Charlotte Islands, 98
Queen, 52
   structures, 54
Qwerty keyboard, 33, 143

Racist, ix, vii
Raids, 95
Rambo, A.T., 20
Ramesside period, 65
Random evolution by mutation and drift, 146
Rank, 96, 97
Rationalization, 179
Raw material, 151, 174
   extraction, 54
   procurement, 37
Real scientists, 230
Recipes, 61, 71, 73, 78
   complex, 61
   cultural, 81
   elementary spatial, 73, 74
   generative, 83
   spatial code, 74
Reciprocity, 94, 150; *see also* Strategy, common reciprocating
Recombination, 142, 145, 152, 179
Redman, C., 21
Red ochre, 202
   stains, 198
Reductionism, 28, 167, 168
Reductionist, 171
Referent/s, 199, 201, 204
Regionalism, 122
Reiff, D., 80
Religion, 45–47; *see also* Institutions
Renfrew, C., 153
Replicatee, 44
Replication, 49, 64
   cultural, 81, 62, 68, 83
   genetic, 81
   multichannel, 81
   rates, 80
Replicative automata, 78
   code, 70
   development sequence, 78
   error recombination, 83
   hierarchy, 78
   strategy, 53
   success, 34, 38

Replicator, 44
   creative, 79
   genetic, 171
Reproduction, 169
   asexual, 230
   cultural, 141
   sexual, 230
Reproductive advantage, 35
   multiobject strategy, 53
   restraints, 94
   success, ix, x, 25, 35, 53, 56, 69, 114–116, 118, 155; *see also* Male
   structure, 51, 53
Residential sector, 77
   space, 80, 81
Resources, 169
   defense, 183
   economic, 100
   human, 96, 174
   lithic, 224
   material, 174
   ownership, 96
   quality locales, 97, 98
   See also Intensification of resource procurement
Retaliation, 150
Reverse transcriptase, 50
Revolution
   Darwin's, 228
   economic, 159
   scientific, 223
   social, 177
   stylistic, 159
Rhetoric, 47
Richardson, J., 137
Richerson, P.J., X, 7, 8, 91, 110–112, 114, 126–128, 143, 150, 151, 154, 155, 157, 170
Rindos, D., 6, 26, 43, 45
Ring, B, 187
Ritual, 48–50
   communities, 148
   dances, 90
   initiation, 148
   knowledge, 148
Rivalry, 231
RNA retroviruses, 48, 49
   -based species, 51
Rock art site, 199
Rogers, R.S., 189
Rome, 117, 151

Roman Empire, 117, 118, 128
Roman Inquisition, 177
Romanization, 118
Roman Republic, 117
Romans, 151
Rossman, A, 97
Rote-learned motor habits, 137
  social learning, 138
Rousseau, J.J.,47
Rozin, P., 207, 211
Rubel, P.G., 97
Ruhlen, M., 49
Rule/s, 27, 71, 73, 171, 225
  conformist, 151
  for innovation, 144
  formal, 122
  marriage, 97
  nonconformist transmission, 152
  positive frequency-dependent, 151
  preference, 157
  social, 175, 189
  social exchange, 188
  symbolic, 155
  transmission, 144
  See also Decision, Inherited characteristics

Sackett, J.R., 120, 121, 126, 137–139
Sahlins, M.D., 94, 95, 157, 158
Salmon, 96, 125, 207
Salmon, M. and W., 22
Sampling, 145, 146
  biased, 151
  Harvard undergraduate, 188
San, 139
Sanctions, 175
Savannah, 46
Scaffold/ing, 173, 175
  of social development, 176
Scavenger, 46
Scavenging
  opportunistic, 206
  primary, 205
Schull, J., 211
Scale, 149
  effect, 137
Science/s, 45, 46, 13, 209, 231
  biological, 167
  cognitive, 183, 201, 207
  Hempelian, 22, 23
  historical, 28
  life, 18

Science/s (*cont.*)
  natural, 5, 13
  physical, 18, 28
  predictive, law-driven, 22
  social, 13
Scientific
  disciplines, 148
  discourse, 46
  evolution, 135
  explanation, 22, 167
  issues, 222
  method, 21, 225
  paradigms, 24
  vision, 225
  See also Revolution
Seahawks, 122
Sea mammal bone club, 124
Seashells, 198
Seattle, Washington, 122
Secondary value selection, 115
Sedentary, 95, 96
Sedentism, 36, 98
  regional, 224
Seed-beaters, 160
Selection
  catastrophic, 120
  cultural, 25, 27, 112, 114–117, 120, 121, 125–127, 166, 170, 172, 230
  cultural–behavioral, 87
  Darwinian, x
  diversifying, 120, 127
  exchange, 183
  external, 70
  individual, 102
  internal, 70, 75, 83
  intersocietal, 113
  kin, 101
  mechanisms of, 10
  natural, 27, 34, 82, 90, 91, 93, 110–116, 120, 126, 134, 135, 141–143, 145, 149, 150, 156, 157, 165–167, 170, 171, 200
  negative, 118, 135, 145
  objects of, 170
  positive, 118, 119, 135
  sexual, 134, 135, 142–144, 227
  social, 134
  stabilizing, 120, 127
  subjects of, 170
  See also Agent/s, Anthropomorphic concept of selection, Artificial selection, Cultural, Directional selection, Group/s, Mate

INDEX

Selection according to consequences, 115
Selection and drift, 26, 27
Selectionist
  cultural, 175
  framework, 38
  theory, 34, 35, 38, 51
  perspective, 39
  metaphysics, 47
  See also Approach, cultural-evolutionary, Culture, Hypothesis
Selective, advantage, 201
  agents, 35, 38–40
  breeding, 170
  environment, 38
  external pressures, 71, 82
  forces, 39
  internal pressures, 70
  metaphysics, 43
  -self, 113
  See also Context, Hunting
Self-interest, 93, 227
  -sufficient, 198
  -sufficient marks, 202
Selfish individualism, 3
Semantic, 153, 209
  content, 202
Semin, G.R., 189
Semiotics, 64
Seriation, 136, 158
  of changing corn frequencies, 148
  of changing gravestone designs, 148
  stylistic, 148
Service, E.R., 92, 95
Settlement/s, of artisans, 62
  change, 224
  complex, 71
  location of, 160
  pattern, 189
  small-scale, 80
  space, 73
  spatial format, 79
  See also Australian Aboriginal settlements
Sexist, vii, ix
Sexual, versus social intercourse, 110
Seyfarth, R.M., 202, 203, 207, 210, 211
Shaman, 47, 148
Sheep, 46, 54
Shennan, S., 128
Shepherd, 46, 54
Showy-tail genes, 142

Sierra Leone Creoles, 154
Signal assemblage, 62
  cultural, 65, 82
  human cultural, 64
  internal sorting, 71
  material, 62
  of a message, 61
  spatial, 65
  surplus, 70
  vocal, 187
Silver, 125
Simon, H.A., 174, 176
Simulation, 226
Sinha, C., 172
Skeena River, 100
Skeletal part representation, 202
Skeleto-muscular apparatus, 186
Slap, J.K., 49
Slaves, 117, 171
Slavery, institutionalized, 96
Smuts, B.B., 192
Social, action, 80
  activity, 213
  aggregate, 91
  archaeology, 62
  behavior, 113
  change, 177
  construction, 175
  contract, 175, 188
  control, 100
  convention, 175
  -democratic, 47
  distance, 94
  environment, 12, 188, 189, 192; *see also* Normal social environment
  field of action, 117
  goal, 231
  hierarchies, 92, 98, 169
  history, 111
  insects, 52, 53, 228
  interactions, 150, 154, 185, 201, 203
  knowledge, 173, 209, 210
  life, 82, 150
  matrix, 177
  meaning, 62
  organization, 5, 80, 160, 229
  personhood, 127
  practices, 176
  products, 173
  rank, 93, 100, 102

Social, action (cont.)
　recontruction, 68, 80
　relationships, 209
　scaffolding, 174
　solidarity, 154
　sphere, 200, 201
　status, 78, 94, 95
　strategies, 12
　theory, 64, 82
　See also Agent/s, Context, Cooperation, Determinism, Intelligence, Isolation, Labor, Revolution, Rule/s, Structure, deep message, Transmission
Socialism, 47
Socially or symbolically motivated behavior, 137
Societies, complex, 113
　industrialized, 116
　large-scale, 100
　small-scale, 101
　stratified, 102
　third world, 116
　village-based, 102
Sociobiology, 25, 47, 87, 94, 111, 171, 193
　ecological niche, 190
　linguists, 154
Software, 150
Soldiers, 52
Somalia, 52
Sorting, 35, 38, 39, 40, 75
　stochastically, 135
Southeast Alaska, 96
Southern Blacks, 154
Southern British Columbia, 124
Space and time, 83, 110, 178, 225
　displacement, 199, 203
Spatial and temporal frame of reference, 68
　scales, 69
Spatial autocorrelation, 151
　dimensions of reality, 81
Spencer, H., 90
Specialist carvers, 126
　full-time versus part-time, 126
Speciation, 27
Stable until pushed, 7, 87
Starvation, 97
State, 87, 94, 100, 102
　formation, 159
　-level societies, 90
　rise of, 100

Statistical association, 152, 153
　drift, 153
　method, 74
Status, 157, 183
　competition for, 100
　defense of, 101
　high, 96
　lineage-based, 101
　striving for, 101
　symbolically defined, 155
　See also Social, action
Steele, J., 8, 188, 190, 192, 193
Stephan, H., 191, 192
Sterile workers, 56
Steward, J.H., 20, 23, 25, 26, 136
Stochastic, 34, 136
　processes, 38
　style, 23
　steady, 169
Storage, 55, 98, 125; see also Plant/s
Strategy, common reciprocating, 150
　content-independent propositional reasoning, 188
　hunting or scavenging, 205
　mass hunting, 206
　plant-intensive Numic, 160
　rare reciprocating, 150
Stratification, 94, 126
Strawman, xiii, 222
Stress, residential, 80
　resource, 93, 100
　social, 70
Stewart, I., 75
Stringer, C., 134
Stone, 125
　tool manufacture, 174
　working, 213
Stonehenge, 176
Storytelling, 26
Structural-functional, 154
Structuralists, 122, 139, 159
Structure, deep message, 79
　internal code, 61
　fractal, 75
　language, 68
　nonverbal meaning, 83
　social, 52, 128
Struggle, 165, 175
　concept of, 169
　for existence, 169, 170, 173–176, 178

INDEX

Struggle (cont.)
  Malthusian dimensions of, 169
  process of, 160
  for survival, 169
Style, 7, 110, 112, 119, 120
  animal, 134
  assertive, 138, 155, 156
  emblematic, 138
  expressive art, 155
  iconological, 138
  isochrestic, 121, 126
  -marked boundaries, 160
  northern, 122, 125
  regional, 122, 127
  sense of, 156
  southern, 122
  versus function, 126
  X-ray, 200
  See also Dual inheritance theory of style, Plant/s, Stochastic
Style and function, 8, 87, 118, 127, 133, 134, 135, 137
  dichotomy, 134, 135, 144, 148, 158
  distinction, 136
Stylistic and functional variables, 155
Stylistic behavior, 134, 135
  diffusion, 147
  elements, 160
  heterogeneity, 144
  turnover, 147
  See also Evolution, Innovations, Revolution, Seriation, Trait/s
Stylistic variation, 119, 140, 153, 158, 159
  iconologic, 121
  symbolic, 121
  See also Chance
Stylistic-neutral patterns of variation, 152
Subassemblies, 176
Subject, 166
Suboptimal genotype, 145
  locus, 145
  troughs, 147
  See also Tool/s
Subpopulation size, 147
Subsistence, 224
  activity, 201, 206
  behavior, 225
  economies, 149, 150
  marine-based, 96
  practices, 5, 183

Subsistence (cont.)
  stress, 101
  technique, 154
  technologies, 134
  See also Optimal behavior, Trait/s
Sugden, R., 150
Sui generis, 229
Suite of actions, 78
  signals, 78
Superorganic, 90
Survival of the fittest, 25
Sutton, D., 186
Swanscombe, 205
Symbiotes, 48
Symbiotic domestication, 45
  ecological relationship, 45
  relationship, 54
Symbolic behaviors, 154, 200
  communication system, 153
  ecology, 46
  encoding, 111
  marker, 155, 156
  meaning of a style, 120
  resonance, 53
  structures, 46
  variation, 137, 156
  See also Human, Rule/s, Status
Symbolism, 8
  capacity for visual, 206
  See also Visual images
Symbols, 112
  create, 213
  emblematic, 139
  expresssive, 139
  human, 204, 205
  noniconic, 198
  nonintentional, 205
  -reading, 201, 206, 213
  visual, 199, 200
Syntax, 153, 209
Synthesis, 112, 126
Systems, biological, 67, 75
  common production, 119
  communication, 73, 154
  cultural, 78, 81, 91, 229
  cultural signal, 75
  dynamical theory, 81
  equilibrium, 93
  evolutionary, 144
  genetic, 78

Systems, biological (*cont.*)
  human sensory, 78
  ideological, 155
  of extrasomatic transmission, 141
  of inheritance, 91
  of knowledge, 111
  marriage, 228
  material, message, 68, 81
  mating, 142
  message, 62, 64, 80, 81
  multichannel cultural message, 67 (illus.)
  replicative, 61, 69, 79
  signal, 69, 78, 81
  slow message, 68
  social, 90, 117, 150, 186, 190
  symbolic, 157, 159
  traditional prestige, 123
  See also Economic, Hominid/s, Human, Information, Nonhuman living systems, Prestige
Systemic knowledge, 119
  organization, 111

Tabula rasa, X, 183
Tangled bank, 224–226
Taphonomic investigation, 159
Taste, 156
Tautology, 100
Taxonomy, 83, 144
Taylor, W.W., 20
Teachers, 148
Tebenkof Bay, 99
Technique, formalization and routinization, 137
  manufacturing, 54
  production, 174
  See also Subsistence
Technology aboriginal, 97
  alternative, 150
  biface, 38
  change, 38
  curated, 36
  expedient core, 36, 37
  flake, 38, 39
  formal core, 37
  lithic, 37,39, 40, 224
  mundane, 160
  prehistoric, 205
Teenagers, 154
Tehuacan Valley, 148
Teleology objection, 168

Template, 79
  mental, 112, 120
Temporal cortex, 187
  lobe, 187
Termites, 52, 55
Territory, 96
  mark, 205
Tesik-Tash, 206
Tests of cogency, 222
Textbook writers, 148
Theory
  biological, 169
  culture transmission, 230
  neutral, 145, 152
  optimality, 5
  shifting balance, 142
  social, 64, 82, 169
  See also Coevolutionary theory, Darwinian theory, Evolution, Hegelian theory, Lamarckian, Mendel's theory of inheritance
Time and energy budget, 40
  astronomical, 176
  biological, 172
  scale, 147
  and space, 97, 102, 133, 138, 145, 153, 168, 204
  and space systematics, 19
  stress, 40
Titleholders, 126, 127
Tlingit, 96
Tool/s, 36
  complex, 152
  interpretive, 143
  mathematical, 222
  stone, 70
  suboptimal, 137
  types, 160
Topography, 37
Top site, 63 (illus.), 65, 73, 75, 76 (illus.), 77 (illus.)
Tooby, J., 102, 134, 207, 210
Torralba, 205
Torrence, R., 6
To select, 166, 170
Totem poles, 122
Tourists, 154
Tower of Babel evolution, 154
Trade, 92, 98
  networks, 153

Traffic speeds, 80
Trait/s
  adaptive, 151
  anatomical, 190
  behavioral, 190
  cultural, 115, 116, 146, 152
  functional, 118, 136, 149
  indicator, 126
  marker, 155, 156
  neutral, 136, 151, 152
  preference, 157
  sexually selected, 134
  stable, 149
  stylistic, 118, 119, 135, 144
  subsistence, 154, 155
Transformation, 18
  cultural, 25
Transmission
  of the art, 126
  biased, 114
  Darwinian model of cultural, 64
  coherent, 112
  conformist, 150–152; see also Conformist transmission bias
  cultural, IX, 7, 11, 87, 111, 112, 126, 137, 138, 140–142, 148, 155, 156, 172; see also Trial-and-error
  forces, 112
  genetic, 24, 138, 148
  ideational, 141
  indirect, 126
  indirect bias, 114, 126
  of information, 87
  neutral, 152
  one-to-many, 148
  social, 49, 111, 112, 115, 117, 119
  unit of, 112
  vertical/oblique/horizontal, 230
  See also Conformist transmission bias, Frequency-dependent, Ritual, Rule/s
Trend-watching, 152
Trial-and-error
  concept of, 173
  model of cultural transmission, 174
  process, 174
  search, 149
Tribal societies, 95
Tribes, dialect, 113
Trigger, B.G., 80
Trivers, R.L., 93

Tropical forest, 101
Truth/s, 47, 148, 167
  God's, 225
Tsimshian, 98, 99
Tunable parameters, 146
Turelli, M., 157
Tycho, 177
Tylor, 91

Upper Paleolithic, 198, 206, 213
  record of Europe, 190
  transition, 159
Units, cultural, 82
  of cultural evolution, 113
  of cultural replication (UCR), 61, 67, 82
  of culture, 229
  empirical, 18
  ethnic, 137
  kin, 97
  minimal, 113
  nuclear, 190
  one-unit genetics, 68
  of replication, 82
  residence, 66 (illus.), 78
  scale of, 82
  of selection, 9, 113
  social, 113, 133
  of social organization, 94
  theoretical, 18
  See also Information, Political alliances
United States, 7, 113, 116
  south central, 120
  western, 160
Urban sociology, 8
Universal/s, 185, 193, 200
  cultural processes, 117
  Darwinism, VII
  statements, 22
  facts, 23, 28
  referent, 65
  significance, 204
  social processes, 117
University of Washington's Quarternary Laboratory, 124

Values, cultural, 109, 116
  detectable selective, 118
  genetic fitness, 116
  inclusive fitness, 117
  judgement, 78

Values, cultural (*cont.*)
  primary, 109, 113, 114
  secondary, 109, 113–115, 117, 121
Variants, adaptive, 152
  behavioral, 34
  cultural, 115, 150, 155
  genetic, 146
  lucky, 146
  neutral, 139
  northern, 122
  regional, 125
  stylistic, 121, 143, 147, 149
  surviving, 119
  technological, 34
  See also Isochrestic
Variation, 18, 27
  bias, 143; *see also* Effect/s
  cultural, 111, 114
  formal, 120
  functional, 158
  genetic, 145
  guided, 143; *see also* Effect/s
  iconic, 137
  inherited, 141; *see also* Inheritance
  isochrestic, 121, 137, 140
  nonsymbolic isochrestic, 138
  neutral, 149
  random, 144, 146, 151
  See also Stylistic variation, Symbolic behaviors
Vayda, A.P., 91
Vegetation, 37
Vehicles of gene replication, 171
Venus, 111
Vessels, 55
Village-based political units, 95
  coastal, 96
  ethnographic, 99
  headman, 95, 96
  historic, 98
  house-depression, 98
  hunters and gatherers, 96
  multi-kin-group sedentary, 101
  plank house, 125
  sedentary, 100
Viral, 48, 69
  phenomena, 46, 48–51, 54
  phenotype, 49
  replication, 69
Virion, 49

Virology, 51
Viruses, 48, 171, 229, 230
  DNA, 48
  microbial, 48, 51
  susceptibility, 230
Visual images, 200, 201, 203, 205
  punning, 124
  symbolism, 197, 198, 200–202, 207, 211, 213
  See also Cues, Symbols
Vitalism, 28
Vocalizations, 202
Vocal tract remodeling, 192
Vogelherd, 198
Vrba, E.S., 35, 70
Vulva signs, 198
Vygotskii, L.S., 172, 173

Wallace, 222
Warfare, 92, 100, 101, 153, 175, 183
  organized, 96
  See also Hobbesian "war each against all"
Washington, 122, 123
Wason task, 189 (illus.)
Watson, P.J., 20, 22
Waste, 36
Wealth, 12, 101, 155; *see also* Material
Weissner, P., 121, 137–139, 150, 153, 155
Wenke, R.J., 26
Western ideology of objects, 177
  societies, 209
Whallon, R., 212
White dialect of Philadelphia, 154
White, L.A., 5, 20, 23, 25, 26, 140
White, L.L., 75
White, R., 198
Whorf, vii
Willey, G.R., 19
Williams, G.C., 90, 91
Wilson, E.O., vii, 152
Wimsatt, W., 158
Wobst, H.M., 198
Wolf, E., 117
Woman, x, 157
Women, 95, 160
  activities, 39, 40
  high-status, 125
  role of, 62
  social lives of, 80
  with children, 80
  See also Labor

INDEX

Wood, 125
Wooden, 126
Woodland Period pottery, 120
Working memory, 186
Wright, S., 147
WWII, 154
Wynne-Edwards, V.C., 90

!Xo, 150

Yanomamo, 95, 97, 101
Yoffee, N, 116
Young, R.M., 166, 169

Zeitgeist, vii
Zeus, 111
Zone of Proximal Development, 173
Zoomorphic elements, 124
Zuk, M., 143

INTERDISCIPLINARY CONTRIBUTIONS TO ARCHAEOLOGY
Chronological Listing of Volumes

THE PLEISTOCENE OLD WORLD
Regional Perspectives
Edited by Olga Soffer

HOLOCENE HUMAN ECOLOGY IN NORTHEASTERN NORTH AMERICA
Edited by George P. Nicholas

ECOLOGY AND HUMAN ORGANIZATION ON THE GREAT PLAINS
Douglas B. Bamforth

THE INTERPRETATION OF ARCHAEOLOGICAL SPATIAL PATTERNING
Edited by Ellen M. Kroll and T. Douglas Price

HUNTER–GATHERERS
Archaeological and Evolutionary Theory
Robert L. Bettinger

RESOURCES, POWER, AND INTERREGIONAL INTERACTION
Edited by Edward M. Schortman and Patricia A. Urban

POTTERY FUNCTION
A Use-Alteration Perspective
James M. Skibo

SPACE, TIME, AND ARCHAEOLOGICAL LANDSCAPES
Edited by Jacqueline Rossignol and LuAnn Wandsnider

ETHNOHISTORY AND ARCHAEOLOGY
Approaches to Postcontact Change in the Americas
Edited by J. Daniel Rogers and Samuel M. Wilson

THE AMERICAN SOUTHWEST AND MESOAMERICA
Systems of Prehistoric Exchange
Edited by Jonathon E. Ericson and Timothy G. Baugh

FROM KOSTENKI TO CLOVIS
Upper Paleolithic–Paleo-Indian Adaptations
Edited by Olga Soffer and N. D. Praslov

EARLY HUNTER–GATHERERS OF THE CALIFORNIA COAST
Jon M. Erlandson

HOUSES AND HOUSEHOLDS
A Comparative Study
Richard E. Blanton

THE ARCHAEOLOGY OF GENDER
Separating the Spheres in Urban America
Diana diZerega Wall

ORIGINS OF ANATOMICALLY MODERN HUMANS
Edited by Matthew H. Nitecki and Doris V. Nitecki

PREHISTORIC EXCHANGE SYSTEMS IN NORTH AMERICA
Edited by Timothy G. Baugh and Jonathon E. Ericson

STYLE, SOCIETY, AND PERSON
Archaeological and Ethnological Perspectives
Edited by Christopher Carr and Jill E. Neitzel

REGIONAL APPROACHES TO MORTUARY ANALYSIS
Edited by Lane Anderson Beck

DIVERSITY AND COMPLEXITY IN PREHISTORIC MARITIME SOCIETIES
A Gulf of Maine Perspective
Bruce J. Bourque

CHESAPEAKE PREHISTORY
Old Traditions, New Directions
Richard J. Dent, Jr.

PREHISTORIC CULTURAL ECOLOGY AND EVOLUTION
Insights from Southern Jordan
Donald O. Henry

STONE TOOLS
Theoretical Insights into Human Prehistory
Edited by George H. Odell

THE ARCHAEOLOGY OF WEALTH
Consumer Behavior in English America
James G. Gibb

STATISTICS FOR ARCHAEOLOGISTS
A Commonsense Approach
Robert D. Drennan

CASE STUDIES IN ENVIRONMENTAL ARCHAEOLOGY
Edited by Elizabeth J. Reitz, Lee A. Newsom, and Sylvia J. Scudder

HUMANS AT THE END OF THE ICE AGE
The Archaeology of the Pleistocene–Holocene Transition
Edited by Lawrence Guy Straus, Berit Valentin Eriksen, Jon M. Erlandson, and David R. Yesner

VILLAGERS OF THE MAROS
A Portrait of an Early Bronze Age Society
John M. O'Shea

HUNTERS BETWEEN EAST AND WEST
The Paleolithic of Moravia
Jiří Svoboda, Vojen Ložek, and Emanuel Vlček

DARWINIAN ARCHAEOLOGIES
Edited by Herbert Donald Graham Maschner